W9-BWA-775

A PLUME BOOK

I CAN HEAR YOU WHISPER

LYDIA DENWORTH is a former *Newsweek* reporter, London bureau chief at *People*, and professor of journalism at Fordham University. Her work has appeared in the *New York Times*, *Child*, *Redbook*, *Good Housekeeping*, and other publications. She lives with her family in Brooklyn, New York.

Praise for *I Can Hear You Whisper*

"Writing with clarity and style, Denworth serves as a capable guide to a world that few with full hearing are fully aware of. . . . A skilled science translator, Denworth makes decibels, teslas, and brain plasticity understandable to all."
—*The Washington Post*

"In this moving and informative book, former *Newsweek* reporter Denworth recounts her emotional and intellectual quest to help her deaf infant son hear. . . . This is a book that parents, particularly of deaf children, may find indispensable."
—*Publishers Weekly*

"All parents will recognize the moments of both terror and pride that mark the journey; parents of deaf children will garner both information and insights."
—*Kirkus Reviews*

"Eloquently explains how hearing works . . . An excellent book for anyone with deafness in the family or with a desire to better understand how people hear, why hearing loss occurs, and how it is treated." —*Booklist*

"Lydia Denworth has written a beautiful book that combines superb scientific reporting with powerful and deeply enjoyable storytelling. Her quest to acquire every shred of knowledge she can to help her deaf son is an odyssey that all parents who worry about their children (i.e., all parents) can intimately relate to. Her discoveries about the workings of language and the intricacies of brain development will change the way you think about hearing, speaking, and selfhood. And her fascinating exploration of the politics of deaf identity is sure to spark a larger conversation about how we talk about, think about, and treat children with special needs in our time."

—Judith Warner, author of *Perfect Madness: Motherhood in the Age of Anxiety*

"Read this if you have ears or ever interact with humans. What a moving and brilliant tour of the scientific, emotional, and political landscape of hearing impairment. As a reader, I'm grateful to Lydia Denworth. As a writer, I'm jealous."

—David Shenk, author of *The Genius in All of Us* and *The Forgetting*

"Denworth provides a lucid, engaging, and thoughtful description of the science of hearing. If you are interested in hearing, speech, and language— as a parent, educator, clinician, or scientist—this book fills an important gap and is a terrific read. Careful about the science and sensitive to the psychological complexities, Denworth provides a masterful account of the path from ear to the brain, from sounds to words."

—David Poeppel, professor of psychology and neural science, New York University

"Lydia Denworth's beautiful personal account and thorough investigation connect the dots between her son's hearing loss; the essential import of spoken language on the developing brain; and what parents, doctors, and teachers can gain from a deeper understanding of how the mind acquires language."

—Dana Suskind, MD, professor of surgery at the University of Chicago and director of the Thirty Million Words Initiative

"*I Can Hear You Whisper* is both an affecting and searching personal story and a fascinating job of science reporting, specifically the science of audiology—how we hear, why some of us don't, and how an amazing but controversial technology was invented. Lydia Denworth's son Alex, the beautiful boy at the center of the personal story, is lucky to have a mother like her. The rest of us are lucky to have such a perceptive, lucid, and touching book."

—Richard Bernstein, author of *A Girl Named Faithful Plum*

I Can
Hear You
Whisper

An Intimate Journey through the
Science of Sound and Language

Lydia Denworth

A PLUME BOOK

PLUME
Published by the Penguin Group
Penguin Group (USA) LLC
375 Hudson Street
New York, New York 10014

USA | Canada | UK | Ireland | Australia | New Zealand | India | South Africa | China
penguin.com
A Penguin Random House Company

First published in the United States of America by Dutton, a member of Penguin Group
(USA) LLC, 2014
First Plume Printing 2015

℗ REGISTERED TRADEMARK—MARCA REGISTRADA

THE LIBRARY OF CONGRESS HAS CATALOGED THE DUTTON EDITION AS FOLLOWS:
Denworth, Lydia, 1966–
I can hear you whisper : an intimate journey through the science
of sound and language / Lydia Denworth.
pages cm
ISBN 978-0-525-95379-1 (hc.)
ISBN 978-0-14-218186-7 (pbk.)
1. Cochlear implant.   2. Deaf children—Rehabilitation.
3. Language acquisition—Parent participation.   4. Denworth, Lydia, 1966–
5. Justh, Alex—Health.   I. Title.
RF305.D46 2013
617.8'9—dc23
2013032769

Printed in the United States of America
10 9 8 7 6 5 4 3 2 1

Set in Dante MT Std with Requiem
Original hardcover design by Daniel Lagin

Contents

PART ONE

QUIET

I

THE COW AND
THE RED BALLOON

At bedtime, it was never totally quiet in Alex's room. Laughter floated down the hall as my husband, Mark, played with our two older boys. The rumble of cars going by on our Brooklyn street or bits of conversation from passersby on the sidewalk drifted through the third-floor window into the narrow room. As Alex and I cuddled in the creaky old rocker wedged next to the crib, we sometimes hit the wall with a bang. From time to time, the radiator behind us hissed. The heating system was nearly as old as our house.

Even so, it was peaceful. My days were full of the noise and energy of three young boys. By evening, when Mark started in on pillow fights and helicopter spins and generally getting rowdy with Jake and Matthew, I retreated with Alex, the baby, to the slow comfort of our routine. It was my time, maternal and feminine, about books and lullabies and snuggling. He sat on my lap, smelling of baby soap and clean pajamas, and I read a story.

"Goodnight room. Goodnight moon. Goodnight cow jumping over the moon," I read one night when he was about eleven months old.

"Look, Alex, there's the cow going up, up, up over the moon. Can you show me the cow?"

He just looked at the book. We'd been reading this book for months. Hadn't his brothers been able to point out animals, colors, and shapes by now?

"Point to the cow, sweetheart."

Silence.

"There's the cow!!!"

Nothing.

When I got to "Goodnight light and the red balloon," I tried again.

"I see the red balloon. Do you see the red balloon?"

Nothing.

"Can you point to the red balloon?"

I took his small index finger in mine and led it to the balloon. "That's the red balloon."

He grinned and I kissed the top of his head. But the question had taken hold of me: Why can't he point to the cow?

I managed to finish the story. As I did every night, I carried him over to the doorway and let him flip the light switch off. We sat back down in the rocker; I rested his head on my chest and started to sing.

I liked to mix it up. Standards like "My Funny Valentine" followed by children's songs like "Row, Row, Row Your Boat." After we chose "Amazing Grace" for my father's memorial service, I sang those verses with a catch in my voice.

As Alex and I rocked in the semidarkness, with the half-closed door blocking some of the hallway light and the chattering of his brothers, my voice faltered more than usual over ". . . was blind, but now I see."

The cow had me concerned.

At the end of the song, I murmured "I love you" into Alex's ear and laid him in his crib with a kiss.

· · ·

The day after Alex was born, four weeks early, in April 2003, a nurse appeared at my hospital bedside. I remember her blue scrubs and her bun and that, when she came in, I was watching the news reports from Baghdad, where Iraqis were throwing shoes at a statue of Saddam Hussein and people thought we had already won the war. The nurse told me Alex had failed a routine hearing test. Failing is memorable. My second son, Matthew, had had the same test when he was born in the same hospital nineteen months earlier, but I didn't remember it at all. I wondered how you could possibly test a newborn's hearing, but was too bleary and tired to ask.

"His ears are still full of mucus because he was early," the nurse explained. "That's probably all it is."

I was given a phone number to call and instructions to make an appointment with the hospital audiologist in four weeks, when the mucus would have cleared. No one seemed unduly worried.

A few weeks later, I strapped Alex into the baby carrier I wore on my chest and trekked from Brooklyn back to Mount Sinai Hospital in Manhattan for his second test. In 1978, a man named David Kemp discovered that when the hair cells in the snail-shaped inner ear react to sound and send signals to the brain, they also send nearly inaudible sounds back up the ear canal—almost like an echo. Those sounds are called an otoacoustic emission (or OAE) and they were what the Mount Sinai audiologist was going to try to measure in Alex. As I cradled his tiny one-month-old body, she put a foam-covered probe in each ear and sent clicks and tones down the ear canals. An OAE test ought to pick up anything worse than mild hearing loss. The results came back normal. Alex's inner ears appeared to respond to the sounds.

"It was the mucus," the audiologist told me. "Everything looks okay now."

Relieved, I headed back to Brooklyn and put hearing out of my mind.

• • •

Having a third boy makes life more chaotic. You no longer have enough arms or a big enough lap. The soundtrack of the days would rival the noisiest dance club. A tolerance for a little danger and dirt helps, as does relaxing about the small stuff. If your first child drops her bottle on the ground, you re-sterilize it. For the second, you rinse it off under running water. For the third . . . well, maybe you pop it in your own mouth to clean it off if water's not available.

Alex made it easy for me. For months, we reveled in what a sweet and peaceful baby he was. "He's taken a look around and figured out he's the third child," his father, Mark, joked. "It's a strategy that's sure to win us over." Perched in his baby seat, Alex took in the world through big brown eyes, always observing. When he got bored or tired, he found his thumb and dropped off to sleep. Even though I knew thumb-sucking would be a terrible habit to break—and it was— at the time I was grateful that Alex could comfort himself so easily. It meant I didn't have to do it. Once, when we had guests, he lay snuggled in a corner of the living room for an hour, quiet as a Buddha, before the visitors even realized he was in the room.

Sometime between six and eight months, however, I was driven out of my third-child complacency. I dusted off my old copy of *What to Expect the First Year* and reread the should-be-able-to's that I had once known by heart.

"By six months, your baby . . . **should be able to** . . . keep head level with body when pulled to sitting."

Not really.

Under the headline "NOT SITTING YET," in the seventh month, I found an attempt at calming anxious mothers. "Because normal babies accomplish different developmental feats at different ages, there's a wide range of 'normal' for every milestone. . . . And since your child has a long way to go before she reaches the outer limits of [the normal] range, you certainly don't have to worry. . . ."

I began to anyway. Alex was slow to hit every milestone: rolling over, sitting up, pulling up, etc. Every time our worry went from simmer to boil, however, he finally achieved the next skill. At his nine-month checkup, when he wasn't pulling up to stand or even holding himself up on all fours, our pediatrician shared my concern. We agreed I would return in six weeks rather than three months if nothing had changed. But again, soon after that, I found Alex standing up in his crib one morning. He had pulled himself up on the railings and was proud as could be. "He's still in the normal range," Dr. Price said when I called to report. "Just the far end of it."

Really, my initial concerns about Alex were pretty low on the scale—a vague uneasiness like the tickle in your throat before a cold comes on that might go away after a good night's sleep.

His first birthday came and went. He liked music. Sometimes he didn't answer to his name, but sometimes he did. Could he not hear me or was he not listening? Could something else be going on?

"The average baby can be expected to say what she means and mean what she says for the first time anywhere between ten and fourteen months. . . ."

Alex could say "hi," "bye," and "mama," but that was it. My other sons both walked at eleven months and started talking around the same time. Then they added more than a word a week, and by eighteen months, they had vocabularies of about fifty words, much like many other children. Alex was clearly different.

"He's the third," people said. "He doesn't need to talk. His brothers talk for him." That argument didn't ring true to me. If anything, I thought having two older brothers would expose him to more language, not less. Apparently, however, my chorus of advisors was not entirely wrong. According to speech researchers, second- and third-born children are generally slightly later talkers than firstborns, and boys talk a little later than girls. I still didn't think that was what was going on.

"Einstein didn't talk until he was three," people reminded me. Maybe. But the chances that I was raising another Einstein seemed pretty slim.

"You worry too much."

I wanted to believe them. My competing instincts met in a stalemate, alternately grasping at any reassurance and needing desperately to know if there was a problem.

Around fifteen months, Alex still wasn't walking or talking. At his regular checkup, the pediatrician, who was someone in the practice I did not normally see, tried to reassure me. She thought he was fine. Afterward, I called our regular doctor. Alex's case was far from clearcut. Plenty of kids are slow to talk; some even wait until they are two but turn out fine. The doctor was wary of sending a child for an evaluation that wasn't warranted. We talked through the possibilities, and by the end of the conversation he agreed to send Alex for an early intervention assessment, the first step in New York City toward getting therapeutic services for a child who is delayed in any way. "If they say there's no problem," Dr. Price said, "that's a good thing."

Our evaluation was set for mid-August at a child development center in Midwood, a Brooklyn neighborhood a few miles to the southeast of us. We drove along Flatbush Avenue past furniture and appliance stores and Dunkin' Donuts outlets. At a traffic light, I noticed a bird on a telephone wire.

"Look, Alex. Bird," I said, pointing out the window.

He looked where I was pointing and said: "Guh."

It suddenly came to me that not only wasn't he talking, he wasn't imitating the sounds of letters correctly. The "buh" in "bird" was coming out as "guh." If I said "mmm," he might say "duh."

The evaluation was designed to assess Alex's gross and fine motor skills, his cognitive skills, and his speech. How was he doing compared to other children his age? Two days before the appointment, he had finally taken a few steps. Now he shakily repeated that for the physical therapist.

For the cognitive testing, we sat on the floor and the psychologist set out some toys. We watched Alex try to figure out what to do with a pop-up toy.

"Does he have any cause-and-effect toys at home?" he asked.

"Um . . . Well, he has plenty of . . . toys," I answered lamely. What the hell is a cause-and-effect toy? I was thinking. Then I studied the toy in question, which had four different buttons to push—a star, circle, square, and triangle. Each one made a different character pop out of the box. It looked familiar.

"We used to have one of these," I added, remembering that my oldest child had had the very same toy at some point. It had broken and been thrown out and was never replaced. Our jokes about third-child syndrome suddenly weren't that funny. If Alex had been the first, of course, I'd have made sure that I knew what cause-and-effect toys were and that he had some.

But cause-and-effect toys were not what he was missing.

The therapists determined that Alex was mildly delayed in his gross motor skills and his cognitive skills. How significant was that? I didn't know. We still had no idea what lay behind the delays, but at least we could take some action. He was eligible for therapy in both areas. However, the area I was most worried about—speech—didn't qualify. A vocabulary of three words was borderline acceptable for a sixteen-month-old and not yet cause for concern. They did recommend another hearing test at eighteen months if nothing had changed.

We began to have visits at home from a cognitive therapist named Linda and a physical therapist named Bernard. Bernard worked on walking, kicking, and jumping. Linda brought stacking games, puzzles, and all the cause-and-effect toys a boy could want. Alex's physical skills got better and better, and he could manage Linda's toys just fine. "There's nothing cognitively wrong with this child," Linda said after several weeks. Still, Alex didn't say another word.

It was time for a hearing test. In early October, we went to an ear, nose, and throat doctor. Sitting in the testing booth, I held eighteen-month-old Alex on my lap and tried to follow the instructions to stay still and quiet. From the adjacent room, on the other side of the sound-proof glass, the audiologist played a series of sounds through the speakers to our left and right. To me, they sounded like static, whistles, beeps, and the low brushing of a whisk broom over a brick floor. The idea was that if Alex heard the sound—each represented a different frequency—he would look at the speaker it came from. To encourage this, when he turned his head in the right direction, an electronic toy lit up. On one side, a slightly demonic-looking monkey played the drums; on the other, an elephant clanged cymbals.

Sounds rolled from the speakers through the small booth, first soft, then louder. I could hear them all. But Alex's little brown head stayed distressingly steady in front of me as he watched the audiologist through the glass. Finally, when the noise got truly loud, he turned his head.

He failed unequivocally. But when we went into the doctor's office and he examined Alex's ears, they were full of fluid and slightly infected. Alex wasn't complaining, but the fluid certainly could explain the test results.

"Let's let the fluid clear up and we'll redo the test," said Dr. Dolitsky. The part of me that wanted to believe the best had found a life raft; the part that needed to know the worst got more frustrated by the lack of clarity. The fluid was stubborn, and appointments with the busy doctor were a rare commodity. Weeks passed.

On a night like many others that cold January, I was making dinner while the boys played in and around the kitchen. I heard my husband's key in the lock. Jake and Matthew tore down the long, narrow hall toward the door. "Daddy! Daddy! Daddy!" they cried and flung themselves at Mark before he was all the way inside.

I turned and looked at Alex. He was still sitting on the kitchen floor, his back to the door, fully engaged in rolling a toy truck into a tower of blocks. A raw, sharp ache hit my gut. Taking a deep breath, I bent down, tapped Alex on the shoulder and, when he looked up, pointed at the pandemonium down the hall. His gaze followed my finger. When he spotted Mark, he leapt up and raced into his arms.

"It's the fluid," I said to myself. "It's the fluid."

The next day, when I went to pick him up from day care, it happened again. I could see him across the room as I came through the gate.

"Hi, Alex," I called. "Hey, buddy, it's Mommy."

He didn't look up or stop playing. The teacher standing next to him said, "Alex, Mommy's here." Nothing. She tapped his shoulder and only then did he look up. His face lit up when he saw me, and he came running to greet me. The teacher and I looked at each other, and I looked away, embarrassed by the awkward intimacy of sharing this terrible moment.

He had never been this obviously unresponsive.

A few days later, on a snowy Sunday, the first in February, I was standing in a Manhattan theater lobby with Mark when my cell phone rang. It was Linda, the cognitive therapist who'd been working with Alex for a few months by then. She had been convinced that hearing wasn't the problem. There were good reasons to think that—Alex had compensated well—and she was not alone in her opinion. "I don't think he can hear," she said now, "at least not well."

I knew that she was right—and that it wasn't just the fluid. Having someone in a position of authority say it out loud allowed me to do the same. The problem now had hard edges, like the difference between worrying about paying the bills and realizing you have to sell your home.

It was twilight as Mark and I left the theater. The snowstorm had momentarily hushed the din of Manhattan. The snow, the buildings, and the sky were washed a dusky blue-gray. Here and there, street-lights sparkled, but their halos blurred and fuzzed before me as I started to cry.

Shock, fear, bewilderment, love. A levee had burst and pent-up emotion flooded through me. Unsteadily, I clung to Mark's arm and blurted out just a few of my frantic questions.

"What will this mean? What will his life be like?" Tears streamed down my cheeks. "How will I talk to my son?" And then, "How could I not have known?"

"If it had been obvious, we would have known," Mark insisted.

"We'll do whatever we have to do," he added with more certainty than I think he felt.

We both knew that most of the doing would fall to me. Mark worked twelve hours a day, longer if he was traveling or had business dinners in the evenings. I had been about to return to work, but those plans were now on hold.

With the next hearing test just days away, I scheduled a speech and language evaluation with a therapist in our neighborhood. Her car-peted office was on the garden floor of a brownstone, which put the small windows at sidewalk level. I sat in a child-size chair and Alex stood at the child-size table. The therapist pulled down some cars and a Mr. Potato Head. An evaluation for a child of twenty-one months con-sists of play and pictures. Out of four pictures of animals, for instance, the therapist might ask a child to "show me the horse." Or, with a line of toy cars on the desk, she might ask the child to "make the green one go." In order to be sure a child is using only hearing, testers sometimes cover the lower half of their faces with a shield, a wooden hoop with a piece of black fabric stretched across it.

When words and instructions were presented to Alex in this way, with the speaker's mouth hidden behind the hoop, it was a disaster.

Until that moment, I hadn't realized just how many visual cues we'd been providing him. Every time we told him to wave bye-bye, for instance, we waved our own hands. As the therapist worked her way through the evaluation, Alex became more and more confused. He wasn't a child who fussed or acted out. He just went still. Quiet and staring, he searched her face and mine for some hint of what was being asked of him. I tried for a reassuring smile, then turned my gaze to the little window to hide the emotion welling up. A stroller went by followed by a pair of black boots. "Do you want some juice?" I heard the woman say to the child.

What stayed with me from the therapist's report of that day were the measures of how Alex was doing relative to other children his age. He was in the second and eighth percentiles respectively for what he could understand (receptive language) and what he could say (expressive language). Overall, 97 percent of almost-two-year-olds had more language than he did. The only good news was that a substantial hearing loss would explain everything he couldn't do and how a child who seemed relatively bright—as measured by the way he played with toys and observed what was going on around him—could have such gaping holes in his abilities and knowledge. It explained why he couldn't point to the cow in *Goodnight Moon*.

The songs I sang at bedtime had been silent for him. Every night I had whispered in his ear "I love you." I don't think he ever heard me.

"I need to know everything there is to know about ears," I told myself at first.

It wasn't that simple. Hearing, through sound, I soon realized, is inextricably linked to spoken language and through language to literacy. I set out to understand ears, but I ended up exploring the brain. I wanted to know what hearing meant for Alex, and found myself grappling with the scientific, developmental, and cultural implications of how to use what we know about sound and language. Alex's ears, in every sense, were just the beginning.

2

A NEW WORLD

I had never met a young person who was deaf or hard of hearing. At least so far as I knew. I was disconcerted by my ignorance but not alone in it. Two or three in a thousand babies are born deaf and one in two hundred loses some hearing by the age of three. More than 95 percent of such babies have hearing parents. For many of those parents, as for me, their own child is the first deaf or hard-of-hearing young person they have ever known.

As word of what was going on filtered out to family and friends, I got a call one winter day from a distant acquaintance. She had a daughter with hearing loss who was in middle school in Manhattan. We spoke for an hour and a half.

"What's his audiogram?" Karen asked.

I hesitated. "I don't know yet." Then I confessed I didn't really know what an audiogram was. She explained that it was the graph of Alex's hearing that his audiologist would create.

"Everything depends on the audiogram," she said.

Then we talked about speech therapy.

"Have you considered option schools?" Karen asked.

Another pause. "What's an option school?" I asked. In New York

City, she said, there were now specialized preschools for children with hearing loss. These were known as option schools.

And so it went. I responded to every one of her questions with two or three of my own. At first, Karen's information felt enormously helpful. Then, almost mid-sentence, it was overwhelming. I stopped taking notes, lay back on the couch, and stared at the ceiling as her suggestions swirled in my head. Which school. What to say to teachers. What to say to other parents. Which kind of speech therapy. What health insurance covered. What it didn't. What early intervention provided. What it didn't. Psychologists and social events for kids and families. Bus transportation. (Bus transportation! Alex wasn't even two yet, but the specialized schools were far away.)

Karen must have guessed what was happening at my end of the line, because she soon said, "Why don't I send you some books I have and you can call or e-mail me with any other questions after you've read them."

I thanked her and hung up. A box of books and pamphlets arrived the next week. Karen and her husband had chosen a speaking and listening approach for their daughter, and the material she sent reflected that—books with titles like *Early Language* and *The Words They Need*. When I'd asked her about sign language, she told me that the view of many in the oral deaf world was that American Sign Language (ASL) delayed the acquisition of spoken language. I would find out just how controversial that idea was.

For those born into deaf families, the experience is radically different. Sam Supalla, a storyteller and filmmaker who is deaf, had deaf parents and two deaf brothers as well as a brother who was hard of hearing. In a book about deaf experience, I found Supalla's description of an early friendship with a little girl who lived next door. They were about the same age and began to play together occasionally. Sam enjoyed her company, but there was the problem of her "strangeness." She didn't seem to understand him at all the way his family did, not even the crudest gesture. "After a few futile attempts to converse, he gave up and instead pointed when he wanted something, or simply dragged her along with him if he wanted to go somewhere," I read. "One day, Sam remembers vividly, he finally understood that his friend was indeed odd. They were playing in her home, when suddenly her mother walked up to them and animatedly began to move her mouth. As if by magic, the girl picked up a dollhouse and moved it to another place. Sam was mystified and went home to ask his mother about exactly what kind of affliction the girl next door had. His mother explained that she was hearing and because of this did not know how to sign; instead she and her mother talk: They move their mouths to communicate with each other. Sam then asked if this girl and her family were the only ones 'like that.' His mother explained that no, in fact, nearly everyone else was like the neighbors. It was his own family that was unusual."

So what did I know about this unusual world? Most of it could be reduced to a few cultural references: *Children of a Lesser God*, Dr. Benton's son on *ER*, Heather Whitestone as Miss America, Marlee Matlin on *The West Wing*. Beyond that, I had a sense of the deaf community's pride. I remembered news articles about student protests at Gallaudet, the deaf university in Washington, DC. I also knew there was controversy over cochlear implants, the revolutionary devices that enabled the deaf to hear. Here again, my knowledge came mainly from a film, the Oscar-nominated documentary *Sound and Fury*, which had been released a few years earlier. It followed the painful debate within the Artinian family over whether to get the prostheses. One brother and his wife—Peter and Nita, both deaf—were leery of the implant. After much soul-searching and exploration, they chose not to get one for their six-year-old daughter, although she wanted one, or for their two younger children. Peter's brother, Chris, and his wife, Mari—both hearing—chose an implant for their son Peter when he was born deaf. Peter and Chris's parents are hearing. Mari's are deaf. The hearing grandparents, particularly Peter and Chris's mother, pushed hard for the implant and were devastated by Peter and Nita's decision against it. Mari's deaf parents were equally distraught over her decision for the implant. The wrenching story of this divided family moved me, but it seemed far removed from my life.

I wasn't facing exactly the same decisions even now. We didn't yet know how much Alex could hear. I did know a few movies and newspaper articles weren't much to go on. So I started reading. Whenever I could find a quiet moment during the day or after the boys were in bed, I planted myself at my computer in our second-floor study and searched the Internet. I was looking for information, of course, but also for something more. Perhaps for comfort. My newly created file, marked "Alex—Hearing," grew fat with printouts from the American Speech-Language-Hearing Association, the League for the Hard of Hearing, the National Association of the Deaf, the National Institute on Deafness and Other Communication Disorders, and any hospital or advocacy group I could find with something else to tell me. On a pad of paper next to my keyboard, I scribbled questions.

"What is an ABR?" I wrote on the night I focused on hearing tests. Under that, I noted "bone conduction?"

The next day, I was fixated on assistive devices. "What is an FM?" I wrote. "What do hearing aids for a little guy look like?"

"ASL—where?" I wrote on another night.

Quickly, I learned just how much variety the terms "deaf" and "hard of hearing" encompassed. Audiologically speaking, hearing loss could be categorized as mild, moderate, severe, or profound, but its specifics, measured in thresholds and frequencies, were nearly as individual as fingerprints. To divide the world into deaf and hearing seemed like calling a thoroughly mixed-race society like Trinidad or Brazil simply black and white.

However much a person could hear, there was another, more

personal distinction: how that person chose to be identified. You could be deaf or Deaf. A small "d" referred to the audiological condition of limited hearing; the big "D" indicated someone who was part of a group that shared a language—American Sign Language in the United States—and a culture. The distinction reached far beyond spelling; it was the difference between thinking about deafness through a medical model or a social model, the difference, as Andrew Solomon later described it in *Far from the Tree*, between illness and identity. There was a long list of terms and ideas to run through this filter. "Hard of hearing," which I initially thought old-fashioned, was the preferred term within the Deaf community for someone with residual hearing. "Hearing impaired" was commonly used in medical pamphlets and some government literature, but the National Association of the Deaf (NAD), the primary advocacy group for ASL and Deaf culture, viewed the term as "well-meaning" but "negative" for its emphasis on what a person *cannot* do. On the same grounds, they objected to the term "hearing loss," although it can be hard to talk about audiology without using that term, I find. You need a noun. Another common distinction concerned the age of onset of deafness. Adults who lose their hearing are "late-deafened," indicating an early life lived in the hearing world. And then there was "oral deaf," the category of deaf and hard-of-hearing people who chose to communicate through "listening and spoken language." They had an advocacy group of their own, the Alexander Graham Bell Association (known as AG Bell)—an antagonist, it seemed, of the NAD.

Later, when I got to know some deaf adults, they repeatedly told me, "There's nothing 'wrong' with Alex." It was a thought-provoking and startling statement, as if I saw the sky as blue and they saw it as green. As soon as they said it, I saw the dangers of the word "wrong," its potential to wound. I hadn't meant it as a broad pathologizing of my precious son but very specifically that there was something in the way Alex's ears worked that prevented him from hearing and that might prevent him from learning to talk. And I meant that that fact worried me. Deaf culture used a completely different vocabulary to describe the same set of facts. Or maybe it was describing a different set of facts. Our perceptions are based on our own experiences. I once read about a tribe in Namibia whose members distinguish between shades of green more easily than between green and blue. Could it be that in a hearing person's view the sky of deafness appears to be blue and for Deaf that sky looks green?

"For hearing people, the world becomes known through sound," wrote Carol Padden and Tom Humphries, professors of communication at the University of California, San Diego, who are both deaf. "Sound is a comfortable and familiar means of orienting oneself to the world. And its loss disrupts the way the world can be known." Deaf people, say Padden and Humphries, have a different center.

They do, however, occasionally turn sound on hearing people, as in this classic deaf joke:

> A Deaf couple check into a motel. They retire early. In the middle of the night, the wife wakes her husband, complaining of a headache, and asks him to go to the car and get some aspirin from the glove compartment. Groggy with sleep, he struggles to get up, puts on his robe, and goes out of the room to his car. He finds the aspirin and, with the bottle in hand, he turns toward the motel. But he cannot remember which room is his. After thinking a moment, he returns to the car, places his hand on the horn, holds it down, and waits. Very quickly, the motel rooms light up—all but one. It's his wife's room, of course. He locks up his car and heads toward the room without a light.

The joke is not on the Deaf man, of course. "He knows he can count on hearing people to be extraordinarily attentive to sound—to his gain and their detriment," note the authors. More aggressively, when Deaf people want to insult each other, they can't do much worse than to accuse someone of THINK-HEARING.

But I am hearing. I laughed at the motel joke when I came across it, but I do use sound to orient myself in the world. I couldn't just undo a lifetime of hearing overnight. Nor was I sure I should. Sound allows us to do some remarkable things.

What I did acknowledge was that words and labels have power, and there is also power in the right to choose those labels oneself. That there were so many possibilities within the Deaf world revealed a complicated history that would take time to probe but that was brimming with both the pride and the tension of which I'd been only vaguely aware.

I could see that how deaf and hard-of-hearing children should be educated and, more fundamentally, how they should communicate are critical questions. "Deafness as such is not the affliction," wrote neurologist Oliver Sacks, "affliction enters with the breakdown of communication and language." Without communication, there can be no education. Communication provides contact and connection, a means of coming together. It allows the sharing of ideas and information. "Language really does take us everywhere," neuroscientist Paula Tallal of Rutgers University, who studies children with language impairments, has said. "If we think about what makes us human and makes us able to function differently, ultimately, it is language; first of all language and of course subsequently written language. From the time we're born, our interaction with our parents, our interaction with peers, our interactions with our sense of self are very wrapped up with the language system."

By "language," I had always taken for granted that one meant "speech."

I certainly took it for granted in my parenting. Until Alex forced me to think about it, I hadn't realized just how much I parent with my voice. I soothe and cajole, read and sing, teach and explain, reprimand, and occasionally yell. It's how I hand down whatever wisdom I have, and how I answer questions, drum up interest, urge compassion, and encourage diligence.

Could I do all that for Alex through sign language? If I were fluent in ASL, maybe. A language that Sacks described as "equally suitable for making love or speeches, for flirtation or mathematics" was no doubt also suitable for mothering. But the idea of raising a child in what to me was a foreign language was daunting. If we were to communicate in ASL, being hearing parents was a distinct disadvantage. Mark and I would be starting at the beginning, like babies, only—unlike babies— with brains that were no longer optimized for learning a new language.

As it happened, at that very moment, baby sign language was all the rage among the stay-at-home mothers of Brooklyn, including some of my good friends who thought signing would bolster their hearing children's communication skills and who urged me to try their videos. This did not encourage me. Actually, it annoyed me. Faced with a child who might truly be deaf, I thought the idea of baby sign classes ridiculous and, in our case, of no more use than a knife in a gunfight. If we were going to learn to sign, we were going to have to go all-in. If I could have seen past my own prickly defenses, I might have found something useful, if limited, in those videos. After all, we had to start somewhere. I did use a handful of signs like MORE and MILK for a time, but I didn't investigate further because I got my back up about it. I didn't shout, "You have no fucking idea what I'm dealing with!" I took the brave passive-aggressive approach of ignoring baby sign language instead.

Intellectually, I was interested in true American Sign Language and appreciated that ASL was a fully fledged language that could open a door to a whole new world. Over the years, my appreciation has only grown. But in my early foray into the deaf world, two things gave me pause. As I read, I discovered an aspect of the deaf community that no one is proud of: a long, alarming history of educational underachievement. Even after nearly two hundred years of concentrated effort at educating the deaf in America, the results are indisputably poor. The mean reading level of deaf adults is third or fourth grade. Between the ages of eight and eighteen, deaf and hard-of-hearing students tend to gain only one and a half years of literacy skills. Education and employment statistics are improving, but deaf and hard-of-hearing students remain more likely to drop out of high school than hearing students and less likely to graduate from college. Their earning capacity is, on average, well below that of their hearing peers. Why? Was it the fault of deaf education or of deafness itself? I did know that reading in English meant knowing English.

Equally disturbing was the depth of the divide I perceived between the different factions in the deaf and hard-of-hearing community, which mostly split over spoken versus visual language. Although there seemed to be a history of disagreement, the harshest words and most bitter battles had come in the 1990s with the advent of the cochlear implant. The device sounded momentous and amazing to me, and that was a common reaction for a hearing person. It's human nature to gravitate to ideas that support one's view of the world, and hearing people have a hard time imagining that the deaf wouldn't rather hear. As Steve Parton, the father of one of the first children implanted in the United States once put it, the fact that technology had been invented that could help them do just that seemed "a miracle of biblical proportions."

By the time I was thinking about this, early in 2005, the worst of the enmity had cooled. Nonetheless, clicking around the Internet and reading books and articles, I felt as if I'd entered a city under ceasefire, where the inhabitants had put down their weapons but the unease was still palpable. A few years earlier, the National Association of the Deaf, for instance, had adjusted its official position on cochlear implants to very qualified support of the device as one choice among many. It wasn't hard, however, to find the earlier version, in which they "deplored" the decision of hearing parents to implant their children. In other reports about the controversy, I found cochlear implantation of children described as "genocide" and "child abuse."

No doubt those quotes had made it into the press coverage precisely because they were extreme and, therefore, attention-getting. But child abuse?! Me? What charged waters were we wading into? I just wanted to help my son. It would be some time before I could fathom what lay behind the objections. Like a mother adopting a child from another race, I realized my son might have a cultural identity—should he choose to embrace it—that I could come to appreciate but could never truly share. Yet I felt strongly that our family had a claim on him, too. First and foremost, he belonged to our culture.

We might have to take sides.

3

How Loud Is a Whisper?

lex stood with his nose pressed against the arched plate-glass window watching the traffic go by on Second Avenue. Mark and I stood with him, pointing out the yellow taxis, the size of the trucks, the noise of the horns. Talking to your child is a hard habit to break, even if you know you can't be heard. On this day, we were hoping for clarity. Instead of doing another hearing test in the doctor's office, we'd been sent to the New York Eye and Ear Infirmary, where there is a battalion of audiologists on hand at all times. The door into the waiting room opened and a young woman with a broad, friendly face and long brown hair stood on the threshold. She glanced at the file in her hand and then looked up.

"Alexander? Alexander Justh?"

From the moment she called his name that first time early in February, Jessica O'Gara told me later, she was evaluating Alex. "It starts in the waiting room." Like detectives, audiologists use every possible clue to figure out what's going on with a child. Is he engaged with his parents? Did he turn his head when his name was called? What toy is he playing with? Is he holding a book upside down? Can he turn the pages? Could there be a cognitive delay? Can this child hear?

Jessica and her colleague Tracey Vytlacil ushered us down the hall. Alex was shy and, of course, quiet, but Jessica's energy was infectious.

"Hey, buddy, we need to figure out what's going on with you," she said, "but first I think I have something you will like."

She pulled out a box of stickers, the mainstay of medical offices everywhere, and let Alex pick out a handful. While we went through his history, he occupied himself putting stickers on his shirt and arm and then on my shirt and arm. Like every audiologist and doctor we would see in the years to come, Jessica had a plastic model of the ear on display and some wall charts to boot. Later, these cross-sectioned models became one of Alex's favorite playthings during waits for appointments.

Unlike the eyeball, which is bigger inside the skull than what we see in the face, the many parts of the ear get smaller and more intricate under the surface. In the plastic models, the outer ear, the part that's visible, reminds me of Dumbo's enormous ear compared to the tangle of circles and snaking tubes that make up the middle and inner ears. In her marvelous *A Natural History of the Senses*, Diane Ackerman likened the inner workings of the ear to a "maniacal miniature golf course, with curlicues, branches, roundabouts, relays, levers, hydraulics, and feedback loops." The design may not be streamlined, but it is effective, transforming sound waves into electrical signals the brain can understand. It is also particularly well suited to the human voice; our keenest hearing is usually in the range required to hear speech. That makes evolutionary sense. Prehistoric ears had to contend primarily with human and animal noise and the occasional clap of thunder. Modern noise dates only to the Industrial Revolution and the invention of gunpowder.

When I call Alex's name, I'm pushing air out of my throat and making air molecules vibrate. They bump into the air molecules next to them—how fast and how hard depends on whether I've whispered, yelled, or spoken in a conversational voice. Either way, I've created a sound wave, a form of energy that can move through air, water, metal, or wood, carrying detailed information.

Traveling through the air, the sound waves are acoustic energy. The outer ear is designed to catch that energy in the folds of the earlobe (the pinna). It does a slightly better job with sounds coming from the front. Cats and deer and some other animals have the ability to turn their pinna to a sound like radar dishes, but humans must turn their heads. Once collected, the waves are funneled into the ear canal, which acts as a resonance chamber.

At the eardrum (tympanum), the waves have reached the middle ear. When they hit the eardrum, it vibrates. Those vibrations in the membrane of the eardrum are carried across the little pocket of the middle ear by a set of tiny bones—the smallest in the body—called the ear ossicles, but more commonly known as the hammer (malleus), anvil (incus), and stirrup (stapes). Converting the original acoustic energy to mechanical energy, the hammer hits the anvil, the anvil hits the stirrup, and the stirrup, piston-like, hits a membrane-covered opening called the oval window, which marks a new boundary.

On the far side lies the fluid-filled cochlea, the nautilus-like heart of the inner ear. The vibrations transmitted from the stapes through the oval window send pressure waves through the cochlear fluid; mechanical energy has become hydro energy. Outside, the cochlea is protected by hard, bony walls. Inside, the basilar membrane runs along its length like a ribbon. Thin as cellophane, the basilar membrane is stiff and narrow at one end, broad and flexible at the other. As sound waves wash through, the basilar membrane acts as a frequency analyzer. Higher-pitched sounds, like hissing, excite the stretch of membrane closest to the oval window; lower pitches, like rumbling, stimulate the farther reaches. Like inhabitants of a long curving residential street, specific sounds always come home to the same location, a particular 1.3 millimeters of membrane and the thirteen hundred neurons that live there, representing a "critical band" of frequencies.

Sitting on top of the basilar membrane is the romantically named organ of Corti. Known as the seat of hearing, it holds thousands of hair cells. Quite recently, scientists discovered a distinction between the functions of inner and outer cells. Twelve thousand outer cells, organized in three neat rows, amplify weak sounds and sharpen up tuning. Another four thousand inner hair cells, in one row, take on the work of sending signals to the auditory nerve fibers. Like microscopic glow sticks that light up when you snap them, the tiny stereocilia on each hair cell bend under the pressure of the movement of fluid caused by the sound wave and trigger an electrical impulse that travels up the nerve to the brain.

The most obvious way to assess hearing is to test how loud a sound has to be to be audible—its threshold. Decibels, a logarithmic scale that compares sound intensity levels, were invented in Bell Laboratories, the source of most things sound-related into the mid-twentieth century. Named for Alexander Graham Bell, decibels (dB) provide a means of measuring sound relative to human hearing. Zero decibels doesn't mean that no sound is occurring, only that most people can't hear it. With normal hearing, a person can distinguish everything from the rustling of leaves in a slight breeze (ten decibels) to a jet engine taking off (130 dB). The leaves will be barely noticeable, the airplane intolerable and damage-inducing. Urban street noise registers about eighty decibels; a bedroom at night is closer to thirty. A baby crying can reach a surprising 110 dB, a harmful level with prolonged exposure. (Thank goodness they grow up.) A whisper hovers around thirty decibels.

The number of cycles in each sound wave, whether it rises and falls in tight, narrow bands or loose, languid swells, determines its frequency, which we hear as pitch and measure in hertz (Hz). The range of normal hearing in humans is officially about 20–20,000 Hz. Just how astonishing that range is becomes clearer when you compare it to what we can do with our vision. The visible light spectrum encompasses one octave (a doubling of frequency, so violet light has roughly two times the frequency of red), resulting in 128 noticeably different shades of color (these are literally measured in a unit called "just noticeable difference" or JND), although in fact there are far more variations than the eye can see. Hearing, on the other hand, encompasses nearly ten octaves with five thousand just noticeable differences. What sounds too good to be true usually is. Only small children really hear that much. Hearing sensitivity declines with age, and what most adults hear spans about half the possible range, 50–10,000 Hz.

Somewhere along the way, some part of the sequence of hearing was not working for Alex. Jessica and Tracey were trying to figure out what. Having two audiologists was going to make a difference in the reliability of the results, because testing children takes as much art as science. Kids get cranky and tired. They don't pay attention. If they're very young, they can't tell you what they hear. The solution is for one audiologist to run the test and the other to play with the child. Both watch like hawks for signs of response and compare notes. Although there are more direct physiological tests of hearing, behavioral testing is still essential because it's a measure of the results of the entire system. It's like the difference between seeing if a dishwasher turns on and seeing if the dishes come out clean.

A quick immittance test—a puff of air through the canal—showed that Alex's right ear was still full of fluid. No air, and therefore no sound, was flowing through. His left ear, however, was not fully blocked. By this point, the fluid alone was not enough of an explanation. "Fluid doesn't make them not talk," Jessica explained. "They might be slower or like louder noises," but children won't fail to talk entirely.

We moved on to testing in the booth. With Tracey running the audiometer and Jessica enticing Alex with stuffed caterpillars eating toy fruit and toppling block towers, they worked their way through loud and soft sounds of every frequency. Alex had to feed the caterpillar or add a block every time he heard a sound.

Watching the test was an anxious experience, like watching a child readying to catch a pop fly when the game hangs in the balance, or willing her to perform a difficult piece at a crowded recital. Only more so. You observe and your nerves jangle.

As Karen had explained to me, hearing tests create an audiogram, a graphic representation of what a person can hear, by drawing a line (two, actually, one for each ear) along the threshold of the softest detectable sound. The x-axis charts low to high frequency from left to right. Decibels run down the y-axis, getting louder from top to bottom. (Some audiograms reverse this and put the louder sounds at the top.) There's a complication, though. Sound is deceptive. When we listen, we tend to think we hear just one pitch, but there's more going on than that. Daniel Levitin, in *This Is Your Brain on Music*, uses the analogy of Earth spinning on its axis, traveling around the sun, and moving along with the entire galaxy—all at the same time. When we call a note on the piano middle C, we have identified the lowest recognizable frequency, known as the fundamental frequency, but there are mathematically related frequencies called harmonics or partials sounding above that at the same time. Our brains respond to the group as a whole, and the note sounds coherent to us. Audiologists must get rid of the harmonics and partials to be sure a person is truly hearing at a particular frequency. They do that by creating either pure tones (one frequency) or narrow-band tones, which do exactly what they say—excite only a narrow band of the basilar membrane.

It's not just sounds like leaves and airplanes that vary in frequency. The words we say, even the different vowels and consonants in each word, consist of waves of different frequencies, generally between 500 and 3,000 Hz. The "t" sounds in "tugboat" for instance contain more high-frequency energy than the "b" and the "g," for which most of the energy is concentrated in lower frequencies. In a normal hearing ear, those frequencies correspond to particular points along the basilar membrane from low to high in a system that works like a piano key-board. When low-frequency sounds excite hair cells at the far end of the membrane, the brain gets the message to recognize not just a jazz riff played in the deep tones of a stand-up bass but also the sound "mm." The sound of the letter "f," on the other hand, has the same effect as the top notes on the piano: It stimulates a spot at the end of the membrane nearest the oval window, where the high frequencies are found.

A portion of the top half of the audiogram is known as the speech banana. It's an inverted arc, roughly what you'd get if you placed a banana on its back on the sixty-decibel line, and is typically shown as a shaded crescent. To be able to hear normal speech, a person needs to be able to hear at the frequencies and decibels covered by the banana. An average person engaged in conversation, about four feet away, will have an overall level of sixty decibels. The level falls by six decibels for every doubling of the distance, and it rises by six decibels for every halving of the distance, which is why it's harder to hear someone who is farther away.

By the end of that day, we knew Alex had an underlying hearing loss, but there was still the complication of the fluid. Over the next two weeks, in quick succession, Alex had tubes put in surgically by our ear, nose, and throat specialist, Dr. Jay Dolitsky, to clear remaining fluid. ("It was like jelly," he told us.) Alex had a bone conduction test, which I now understood measures what you hear through your bones. Every time you hum or click your teeth, you hear the resulting sound almost entirely through your bones. When you speak or sing, you hear yourself in two ways: through air conduction and bone con-duction. The recorded sound of your voice sounds unnatural to you because only airborne sound is picked up by the microphone and you are used to hearing both. Finally, Alex had an auditory brain stem response (ABR) test, under sedation, which allowed Jessica to measure his brain's responses to a range of frequencies and intensities and pinpoint his level of loss.

When it was all over, we knew that, in medical terms, Alex had moderate to profound sensorineural hearing loss in both ears. That probably meant that his hair cells were damaged or nonexistent and not sending enough information to the auditory nerve. In someone who is profoundly deaf, who can hear only sounds louder than ninety decibels, almost no sound gets through. In a moderate (40 to 70 dB) or severe (70 to 90 dB) hearing loss, that all-important basilar membrane still functions but not nearly as well. Like a blurry focus on a camera, it can no longer tune frequencies as sharply. The line on Alex's audiogram started out fairly flat in the middle of the chart and then sloped down from left to right like the sand dropping out from under your feet as you wade into the ocean. He could hear at fifty decibels (a flowing stream) in the lower frequencies, but his hearing was worse in the high frequencies, dropping down to ninety decibels. He could make out some conversation, but my whispered "I love yous," at thirty decibels or less, would have been inaudible.

Hearing loss doesn't just make the world quieter, it garbles it. When the ear can't tune sounds as sharply, they blur into one another. To someone with some low-frequency hearing, speech sounds dull and muffled, so that even what is audible is hard to understand.

This explained why it had taken so long to detect Alex's loss. He had been hearing some things but not others, responding to some things but not others. What he did hear, he didn't hear clearly. The fluid had intensified the problem but also made it obvious, allowing us to act. A profoundly deaf child is far easier to identify than one who is hard of hearing. Alex's hearing loss also suggested an explanation for the delays in his gross motor skills as a baby; hearing and balance are both centered in the inner ear. Damage sometimes, though not always, encompasses both systems.

Uncovering Alex's hearing loss had been like falling downstairs in slow motion. It dragged on, with information coming in fits and starts. Now, perhaps, we had come to rest, could catch our breath, take stock, and start climbing back up the steps.

With some usable hearing, and hearing aids, there was every reason to think that Alex could achieve spoken language. When it comes to learning to speak, the difference between those who are profoundly deaf from birth or before learning language and those who are hard of hearing has historically been stark. In one state's survey, only 25 percent of children who started out profoundly deaf were able to speak intelligibly by the age of five or six. The statistic was reversed for those with mild to severe hearing loss: 75 percent could be understood when they talked.

Karen had been right about the audiogram. Our decision hinged on it. Since Alex appeared to be able to hear most speech using hearing aids, and he would be right in the speech banana for everything but the highest frequencies, we decided to make speaking and listening our immediate goal and to learn ASL later as a second language.

Sound became essential, the all-important sensation on which everything depended. Like a musician who trains to play by ear, to identify if a note is flat, to pick out oboes over the clarinets, Alex was going to need to practice, practice, practice, and practice his listening some more. He had a lot of catching up to do. He was nearly two and he could say only "mama," "dada," "hello," and "up."

"We figured out why Alex isn't talking," I explained to Jake and Matty, who were then six and three. "He can't hear very well." Although they'd been caught up in their school lives—in first grade and preschool respectively—they had certainly noticed that Alex and I had been spending an awful lot of time going to the doctor and the audiologist. And they had periodically complained.

Even before that, they hadn't been overly thrilled at the prospect of another baby competing for attention. Matthew had been offended at having his role as youngest in the family usurped. Only nineteen months old when Alex was born, he'd thrown a spectacular tantrum while visiting us in the hospital and then mostly ignored Alex in his first year of life, as if that might just make him go away. Jake had already been toughened up by one new brother's arrival. "Can you be my mommy and daddy be Matty's mommy?" he'd asked me plaintively the night Matthew came home from the hospital. He was more accepting of Alex but still capable of an occasional fit of pique over the time I had to devote to the baby.

So I was relieved and gratified when the big boys reacted to my announcement by dropping to the floor, leaning in close on either side of Alex, and hollering: "We love you, Alex!" Then all three grinned. Alex loved being the center of attention; the others figured they'd just been given license to yell in the house.

4

A STREAM OF SOUND

In any one of the nearly seven thousand languages of the world, babies begin to communicate along a roughly similar schedule. Most begin to talk around their first birthday. They put words together in simple sentences like "eat cookie" at about a year and a half. They pick up as many as ten words a day in their twos and, by the time they're three, most are speaking in sentences and know over one thousand words. For English speakers, there are only another fifty thousand more to learn by adulthood.

How do children do it? Here is where the answers get harder. Learning language is "doubtless the greatest intellectual feat any one of us is ever required to perform," said Leonard Bloomfield, a major linguist of the early twentieth century. It's so difficult, computers still can't do it; so far, they have been successfully programmed to fluently understand only one speaker at a time. Yet babies and young children master this enormous task so naturally, few of us even remember making the effort. From Saint Augustine to Charles Darwin to Noam Chomsky to modern-day linguists, a subset of whom are also neuroscientists, a long line of thinkers have pondered the question. Today's views on how babies accomplish the feat owe much to the ideas that

Chomsky put forward in the late 1950s, when he burst onto the language scene and spearheaded the cognitive revolution in linguistics and psychology.

Up to that point, the behaviorists, led by B. F. Skinner, held sway. Expanding from Pavlov's famous experiments in which dogs could be made to salivate at the ringing of a bell, the behaviorists maintained that animals and children were essentially blank slates and could be conditioned to do almost anything, provided the stimuli and setting were right. Language, argued Skinner, was just another behavior, a "verbal behavior." Chomsky disagreed and wrote a devastating review of Skinner's work.

Chomsky's main idea directly contradicted the behaviorists and was hugely controversial at the time. Some aspects of it are still debated— even by Chomsky himself—but many of its tenets are widely accepted today. He argued that babies arrive in the world with an innate ability for language. Nature, said Chomsky, has provided children with a surprising level of knowledge about language that they can't have had time to learn—"the language instinct," Steven Pinker called it in his bestselling 1994 book of that name. Chomsky believed children had a native ability to deploy what he called "universal grammar," referring not to the details of parsing sentences but rather to an unconscious, tacit sense of some basic universal principles of language—so basic they apply whether the child will grow up to speak English, Swahili, or Chinese. For example: All languages have consonants and vowels, they have nouns and verbs, and they have pitches, contours, and intonations; phrases, not words, are the building blocks of sentences, and the rules governing how one can move those phrases around are the same. Universal grammar explained how children could know that Chomsky's famous nonsensical sentence—"Colorless green ideas sleep furiously"—was grammatically correct, while the same words rearranged—"Furiously sleep ideas green colorless"—created a sentence that was gobbledygook, neither grammatical nor understandable.

Even if the ability to learn language is innate, we do not all begin speaking equally well. Language literature is populated with examples of "wild children," such as Victor of Aveyron, who lived alone in the woods of eighteenth-century France until he was about twelve, or Genie, a California victim of horrific abuse who was discovered in 1970 after she had been kept locked in a bedroom and tied to the furniture for the first fourteen years of her life. These unfortunate children had almost no exposure to language, among other things, and provided an unusual opportunity for study. Neither ever achieved normal language skills. The fact that deaf babies do not automatically learn to talk also tells us that the skill is not purely innate. Yet if a deaf baby's parents are fluent signers, he or she becomes a native user of sign language, which adheres to universal grammar, and the baby will follow the same path to fluency with visual language as hearing babies do with spoken language. So while speech may not be innate, certain patterns of language learning seem to be.

"Language is a super-interesting learning problem," neuroscientist Elissa Newport told me as we sat in her new, somewhat bare office at Georgetown University Medical Center. Newport specializes in the acquisition of language. For twenty-three years, she was at the University of Rochester, the last twelve as chair of the Department of Brain and Cognitive Sciences. When I met her, she had just moved to Washington to head a new center for brain plasticity and recovery, where she is studying how young children who suffer a certain kind of stroke recover their language. Her straightforward, no-nonsense style is evident in everything from her short hair to her scientific approach. "We know that languages of the world have a certain type of organization that you don't see in any other species' communication system, and languages of the world have a lot of interesting profound similarities to one another. So there's a very interesting problem to explain: How did we get languages like that, and how do you learn them, and what kind of brain mechanisms are required to do that?"

When Alex was turning two, I hadn't yet realized how important the question of brain mechanisms would be. I was fixated on the second of Newport's questions: How do you learn language? I sought out Newport because I wanted to understand not just what Alex couldn't do but also what other children could do and why. Following the lead of scientists, I knew I needed to understand what was typical in order to better make sense of what was atypical. What I found was worrisome—for what it was clear Alex didn't get as a baby—but also a little bit reassuring, as I began to appreciate that he had managed to learn some important things about language with very little help from sound.

One fact scientists agree on isn't surprising anymore: To learn a language, it's best to start young. "We certainly learn languages as adults but not to the same degree of proficiency," says Newport, "and there's much more variation among individuals as we get older." To say that children are "better" is too simplistic. "Young children don't really learn faster or better," says Newport. "They learn more slowly. It's kind of tortoise and hare. If you look at people who move to a new country, adults are generally faster, they just don't get as far. They do it differently and they don't end up as good."

Newport gave me an intriguing example. She has spent much of the past ten years making up what she calls "miniature languages." In different studies, the same eight verbs and fifteen nouns carry different meanings. So "kleidum" means "drag" in one instance and "head-butt" in another (one can only imagine the story that language will tell). Words like "tombat," "nagid," and "melnawg" might mean "singer" or "baby carriage" or "shopping cart." Newport teaches these limited strings of invented words to babies, children, and adults in her laboratory using pictures and videos. They are then tested on comprehension and production (naturally, the babies aren't expected to speak). Over the course of as few as five days, Newport uses behavioral tests and brain imaging to watch language acquisition unfold—the whole endeavor is like applying time-lapse photography to learning. (It should be noted that these are all second languages for the subjects.)

On the question of the differences between learning as a child and as an adult, she told me that in a recent series of studies, she added some inconsistencies, or errors, to the miniature languages. Five- and six-year-old children acquired the regular parts of the language, not the errors. Adults reproduced more inconsistencies. This is an example of the less-is-more hypothesis, argues Newport. "Kids are more cognitively limited. For certain kinds of tasks, that may be an advantage. They don't get it all at once. They actually can't acquire all the little irregular things about languages, all the details. They get the more consistent, regular parts. That produces a staged type of learning where you get the big patterns first and then you get the little details much later. Adults get all the details right at the start, and they never get the big patterns."

Her study reminded me of a moment when Matthew, our middle son, was about four—certainly, he had not yet learned to read. Standing in the kitchen one Saturday, Mark posed a riddle he had just read: "Mary's father has five daughters: Nana, Nene, Nini, and Nono. What's the fifth daughter's name?" The name Nunu was on the tip of my tongue, when Matthew piped up. "Mary," he said, and gazed at Mark and me as if nothing could be more obvious. I stared back at him and then realized that he was right.

"How did you know that?" I asked. Then I turned to Mark. "He's brilliant!"

No, I realized later, he was simply four and had not yet learned that if A, E, I, and O are presented in sequence, it's a very good bet that U will follow. When Mark told the joke, I knew too much about vowels, and Matthew, who was a charming, talkative, outgoing child, but not necessarily brilliant, knew just the right amount, which is to say, very little.

"That's exactly the same thing," agreed Newport when I told her the story. Language learning is really a matter of learning patterns. "By mechanisms we don't totally understand, we store an incredible wealth of quantitative details about how sounds combine," says Newport. "That's basically what learning a language is about. You do it with the sounds of your language. You do it with the meanings and what people are referring to. You do it with the sequences in which words occur. You don't just memorize all the sound sequences. You somehow compute. You keep track of things that are very frequent combinations, frequent categories."

For spoken languages, you do all of that by listening. "The first year of life is largely a silent rehearsal," wrote linguist Charles Yang in his book *The Infinite Gift: How Children Learn and Unlearn the Languages of the World*. The process starts before birth. Around six months of gestation, expectant mothers begin to feel the baby kick in response to loud noises. Bathed in amniotic fluid, the fetus can't generally make out words, just as you can't if you put your head underwater, much as you might have tried in the pool as a child. What a baby in utero does hear, over the low-frequency sounds of his mother's blood flowing through her body and her steady heartbeat, is what linguists call prosody, the rhythm and contours of the mother's native language. Babies hear enough of it to recognize and prefer their mother's voice once they are born.

How much there is to listen to in the first months and years of life has been shown to have a powerful effect on language ability or lack thereof. A landmark 1995 study by Betty Hart and Todd Risley showed this most starkly. Hart and Risley had been studying early education for many years, even before Head Start was created. No matter how many creative ways they developed to bolster language in their programs, they were perplexed by the persistent difference in vocabulary growth between middle-class children and poorer children. Field trips, directed experiences, and discussions, all of it was ineffective.

"We began to ask what went on before we ever saw these

children, because they started in preschool at age four," Risley once explained in an interview. A little basic math revealed that if children were in preschool for a total of sixteen hours from Monday to Friday, they were awake for at least another sixty to seventy hours in a week. That begged the question of what happened at home. Hart and Risley decided to look at "what's going on in children before we ever see them in preschool, before they're four, while they're learning to talk," as Risley put it. Over two and a half years, beginning when the children were nine months old, they sent observers into forty-two families in Kansas City for one hour every month to record every word that was said between parent and child. The families represented the entire socioeconomic spectrum from white-collar educated professionals through to parents on welfare.

What they found was so surprising Risley called it a "discovery," not just a result. They had expected that the content of the language was most important. Instead, the most significant and fundamental factor was a massive difference in the amount of talking, the sheer volume of words that some children heard compared to others. The children of educated professional parents tended to hear as many as 2,100 words per hour. Those with uneducated parents on welfare heard 600. The average child heard 1,500 words an hour. Extrapolating those numbers to a year's worth of listening, Hart and Risley calculated that children with the most talkative parents had heard forty-eight million words by the time they were four. Those at the other end of the spectrum had heard a fraction of that, only thirteen million words—a gap of thirty-five million words.

Secondly, Hart and Risley did find what they expected: There were qualitative differences in the language the children heard, and those did matter. All parents used a certain amount of similar baseline vocabulary to instruct and inform children: "Stop that," "Hold out your hands," etc. Subtracting that kind of talk, which Risley calls "the business language" of parenthood, left very little conversation in the quietest homes. But talkative parents engaged in what Risley called "language dancing." "The talkative parents are taking extra turns, responding to what the child just said and did, and elaborating on it, caring."

Together, the amount of words and the quality of the "extra talk" had a direct, strong effect on the child. By the time they were three, children who had heard more words and more interesting talk had higher IQs and larger vocabularies. When Hart and Risley tested the children again at nine, the effect was still clear: The difference in early language exposure accounted for differences in vocabulary size and IQ. In fact, early language exposure canceled out socioeconomics or race. In the cases where a poorer parent talked a lot to a child, the child did fine; and if a professional parent did not talk to a child, the child struggled. As the children moved through elementary school, the differences had academic repercussions.

Clearly, experience matters. Even today, psychologists and linguists continue to recalibrate their assessment of the balance of power between what is innate in language and what is learned. There is a lively, ongoing debate, for instance, on whether it's the ability to use language specifically that's innate—i.e., is language a unique skill that is separate from everything else we do; is it "special"?—or is it that pattern-learning abilities are innate and we apply them to language as we do to other skills? Either way, most argue that the old debate about the relative importance of nature (what is neurologically determined) and nurture (what is culturally determined) is not just moot but even "misguided." In *The Scientist in the Crib*, Alison Gopnik, Patricia Kuhl, and Andrew Meltzoff, who are researchers at the forefront of thinking about what babies can do and why, have this to say: "For human beings, nurture *is* our nature [emphasis original]. The capacity for culture is part of our biology, and the drive to learn is our most important and central instinct. The new developmental research suggests that our unique evolutionary trick, our central adaptation, our greatest weapon in the struggle for survival, is precisely our dazzling ability to learn when we are babies and to teach when we are grown-ups."

Neuroscientists like Elissa Newport focus on the neurobiology of language learning and separate out discrete units for study. A developmental psychologist like Athena Vouloumanos, whose laboratory I visited at New York University, takes a different approach as part of what she calls the "interactivist" school of thought. "Now the shift is looking at language as a human, cooperative, social activity," she told me. "What's the relationship between language competence and abilities and the rest of human social life?"

Evolutionarily speaking, language is the most sophisticated tool that humans have mastered. It can be glorious and evocative or dirty and demeaning and is often described as our crowning achievement as a species. It allows us to tell stories, to instruct and inform, to write novels and reference books, to recount our history. To be shut out of language, as the deaf were before the development of sign language, is to be shut out of culture and society. "Something special is going on with people," says Vouloumanos. "Some part of our evolutionary history has given some genetic underpinning to our language ability. That's an obvious sign that some part of that ability is innate."

Although birds and chimpanzees have been taught to do amazing things, not one animal has yet been able to generate original language. Children do this every day when they produce sentences that they did not hear an adult speak, such as "The sun is sweating you," to take an example from the son of linguist Charles Yang. "Children," points out Yang, "are innovators, not just imitators."

Songbirds interest linguists a lot, however. Though they may never learn to talk, they do learn to sing in much the same way babies learn to speak. Like humans, songbirds are born with an innate ability to generate their songs, but they don't begin singing at birth. First, they listen to their fathers (and in a very few species their mothers) and memorize the songs. At first, their own attempts at sound are really "unstructured chirps," much like a human baby's babbling. With practice, they begin to sound more and more like their parents. Scientists have experimented with songbirds in various ways. Chicks hatched in a laboratory were later exposed to the songs of a variety of species, including their own. They showed a clear preference for their native song. But in other experiments, baby birds that never heard their native songs never learned to sing at all. Furthermore, any exposure to their own songs had to come within a sensitive period when they were still able to learn to produce them. Alex, I would learn, was under a similar deadline.

Babies, unlike birds, cannot be raised in a laboratory. If they are hearing, they are enveloped in sound from birth. One of the first challenges a baby faces is pulling apart the stream of sound that surrounds her every day. I imagine a baby literally carried along in a basket on wave after wave of sound: music playing on a radio, her parents talking, a big brother pretending to be an ambulance, water running, a dog barking, birds singing in the tree outside. Adults effortlessly sort through these competing noises, attending to speech as the most critical element. Before they can do the same, babies have to understand that they should.

In her Infant Cognition and Communication Laboratory at NYU, this is exactly the question Vouloumanos is asking: Is there something special about speech for babies from the start? She works with babies every week, trying to piece together the process of speech perception. "We [humans] understand before we can do," says Vouloumanos. "How do we come to that understanding?" On the day I visited, twelve-month-old Elijah was back for the second time. He had first come to the lab at six months and proved himself a natural. Outgoing and cheerful, he cruised along the bright red couch in the newly renovated reception area to make his way to the basket of toys he spotted as soon as he entered the room. It wasn't hard for the adults in the room to deduce that Elijah wanted to get to the blocks; his eyes were glued to the prize.

But how could we tell what he, in turn, knew about us and our intentions? At what level did Elijah, who couldn't yet say more than a few words, understand the words he heard?

To study babies, researchers have developed habituation studies. They present an infant with a series of the same sound or image until he or she becomes bored. Then they change the sound or image and the baby will indicate that he or she recognized the change. In newborns and very young infants, they do that by sucking faster on a pacifier. Put sensors on the pacifier and you can measure the rate of sucking. In older babies who can focus their attention visually, researchers measure "looking time." A baby will look longer at something interesting or surprising than at the same old, same old.

Elijah was part of a study that day in which Vouloumanos was asking whether children his age understand that language is communicative. That is, whether they expect that a word spoken by one person can affect the actions of another. Elijah sat in a room on the lap of his mother, Jody, and was presented with a little play.

A purple triptych screen like a Punch and Judy puppet show was arranged in front of him, with a window onto a small stage in the middle panel. Graduate students and interns played different roles. In the first scenes, one "actor" tried to put a cube in a box but couldn't reach. Three times, in three different "scenes," the actor tried to reach the box, in order to familiarize Elijah with what was going to happen. This was the habituation phase. Then a second "actor" appeared, one who *could* reach the box. Now the first actor could try to get the second actor to help. In some scenes, that first actor turned toward the second and said a nonsense word—"koba"—followed by the second actor successfully putting the cube in the box. That result makes sense if you expect the spoken word to communicate the first actor's intention. In other scenes, instead of turning and speaking, the first actor cleared her throat. Throat clearing should not communicate anything to actor two except that perhaps actor one is getting over a cold. So if, after the throat clearing, actor two *does* do what actor one wanted, that should be surprising. Elijah should look longer because it should not be the ending he expected.

Alas, on this particular day, even a cooperative baby like Elijah is unable to complete the trial. He gets caught up in trying to make his mother look at the play performed in front of them, and keeps turning his head away from the stage. He doesn't pay enough attention to the action itself. So it goes when you're working with babies.

Over the past ten years, however, Vouloumanos and her colleagues, particularly Janet Werker, her advisor when she was at the University of British Columbia, have pulled enough information out of babies to give us a good idea of the somewhat methodical way babies focus on and master different aspects of sound. Before three months of age, for instance, babies show a preference for speech over other sounds, even if the other sounds are acoustically similar to speech. By six months, a baby knows the difference between speech produced by humans and sounds produced by animals. Vouloumanos tested this by showing babies images of human faces, ducks, and rhesus monkeys while playing human speech, ducks quacking, and monkeys grunting. Babies looked longer at human faces when they were listening to human speech. An interesting additional result: If the humans were not speaking but laughing or making other non-speech sounds, the babies didn't show any preference for the human faces, which suggests that speech is indeed special and distinct from the other noises that come out of human mouths.

By nine months, a baby can tell if the sound sequence he's hearing is part of his native language or not. We know this because there are certain sounds, known as phonemes in linguistics, that appear in one language but not another. At birth, early language expert Patricia Kuhl has shown, babies are multilingual. They arrive in the world able to hear and distinguish the various phonemes of all the world's languages. That's logical, since a baby doesn't know until he is born what his native tongue will be. Over the course of the first year of life, babies hear more and more of the language they will grow up speaking, so those phonemes become familiar. Gradually, as Charles Yang puts it, they "unlearn" the other languages. This explains why a Japanese baby can distinguish between an English "r" and "l," but a Japanese adult cannot without extensive practice.

It also explains why, when researchers played tapes of a person speaking in French to French babies, the babies didn't really care much, so to speak, but when they played someone talking in Russian, the babies took note of the difference. For an English-speaking child, says Vouloumanos, this is like recognizing that "splot" could be an English word, but that "vzglad," which means "looking" or "glaring" in Russian, could not.

At twelve months of age, a baby has progressed to working on meaning. When she hears her mother talk, she looks around to try to see what Mom might be talking about. She is trying to find a "referent" for the words her mother used.

The cues babies use to do all of this fall into three categories: linguistic, acoustic, and perceptual. Linguistic cues depend on having some kind of sense of language sounds—for example, understanding that there is a relationship between "is" and "-ing" in English. If you hear one, you often hear the other. Acoustic cues not surprisingly include what a child hears. An example is the fact that pauses in sentences mark boundaries, not necessarily between words but between phrases. "If a kid is trying to figure out how language works," says Vouloumanos, "chunking out units by phrases lets you analyze them separately."

Perceptual cues generally involve higher-order thinking and include a range of factors from the meanings of words to visual hints. I wondered about visual perceptual cues, since they must have been important to Alex. On the wall of my office, there's a photograph of Alex and me taken the summer he was one, right around the time of his first evaluation. I'm carrying him on my chest in a sling as we walk at the beach. I'm gazing into his face and laughing. He is smiling up at me under his sun hat. It looks as if we'd just shared a lovely joke. I'm sure I said something silly, just as I'm sure he didn't hear me. Despite that, the picture is evidence of communication. From my facial expression and perhaps my gestures, to which he must have been keenly attuned, he knew to start giggling.

Deaf babies learning sign language, Vouloumanos told me, use a fuller range of visual cues that echo the acoustic cues hearing babies hear, such as watching for pauses between phrases. They babble with their hands and when they do vocalize the sounds of English, they use far more sounds that they can see (such as "ba") on the lips of the talker. Blind babies, by comparison, babble differently. They use more sounds that do not rely on visible articulations, such as "ga."

Vouloumanos is in the midst of studying gesture by asking if babies understand from the beginning that pointing is communicative. Her hunch is that they do, that babies understand that someone who is pointing is trying to direct their attention. There is disagreement about how old an infant is before she understands that by pointing at an object or person, I want to tell her something about that object or person. If a baby points at a toy, is he underlining his own interest in that toy? Or is he trying to get his mother to share his interest? When his mother points at a different toy, she's creating a line of sight and movement with something interesting at the end of that line. The studies thus far have shown that by fourteen months of age, babies do understand what they can accomplish by pointing. When we met, Vouloumanos was bringing eleven-month-olds into the lab to see what they knew. What I knew was that during the year he was one, Alex had an entire vocabulary in his forefinger. Point to Mom, point to the swing, point to himself. Mom gets the message that he wants to be pushed. The length and strength of the pointing indicated the intensity of his feelings.

Social cues are particularly intriguing to researchers these days. If, as Vouloumanos suggested, language is a cooperative, social activity, it's important to understand what aspects of it derive from social interaction. The latest research attempts to tease out the relationship between language and social skills. In general, babies with better social skills also develop better language skills. Returning to songbirds for a moment, neuroscientists have found that the neurological period in which young birds are able to learn can be extended by social interaction. For humans, the question is, "Which comes first?" says Vouloumanos. Or to put it in academic terms, which is the "motor of development"?

Finely honed social skills were a good explanation for how Alex functioned in his first two years of life and how he compensated for his inability to hear the words people were saying. In retrospect, it was obvious. At the time, it was subtle. If someone asked him to throw away a piece of crumpled paper, he walked backward toward the garbage can, watching for confirmation that he had guessed correctly what was wanted. Nods and smiles were a pretty good indication he was right. At the child care center he attended a few days a week, the teacher always helped the children wash their hands after playing outside. Armed with a washcloth, she'd ask Alex to hold out his hands. He held out his hands. One day, armed with a washcloth as usual, she noticed a smudge on his cheek and told him she wanted to wash his face. He held out his hands. It was only later, heartsick at having missed the signs, that the teacher remembered the incident and told me about it.

I described these examples to Vouloumanos. With two older siblings and talkative parents, she suggested, "Alex had all the social cues: smiling, interacting, turn taking. The social scaffolding was there. Linguistic input was the thing he wasn't getting. He developed a narrative that didn't have language in it. It was based on his understanding of people as social beings, facial cues, and gestures."

Just how critical social and visual cues, gestures, and facial features were to Alex was painfully apparent at that first speech language evaluation back in January, when they were taken away. I compared what happened there to a moment a few weeks earlier, at Christmas. Jake had received a marble run as a gift from my aunt Nancy. After putting it together with the grown-ups, six-year-old Jake showed a delighted Alex how it worked. He held out a marble for his little brother and pointed to the spot where Alex should place the marble to start the run. Jake was talking and gesturing all the time.

"Here, Alex, you try. It goes right here. Wait, wait . . . There it goes!"

Alex took the offered marble, put it in the right spot, and clapped with delight as it ran through the track we had constructed.

The problem was that there was only so far you could go with the kinds of cues Alex had been using. Life was not all marble runs and helpful brothers. If Alex couldn't eventually parse out the parts of the language that were swirling around him, he would have a hard time ever making use of those parts himself.

5

"Some Means of Instructing"

There is a story that goes like this: On a dark night in Paris in the 1760s, a priest was making his rounds in a wretchedly poor neighborhood. Well into his fifties, white-haired and portly, and wearing the long black cassock of a religious man, the Abbé Charles-Michel de l'Epée traveled along a narrow cobblestone street, through a bleak courtyard, and up a steep, worn stair until he found himself in a dimly lit meager room where two teenage sisters clad in dark wool dresses sat sewing on stools by the hearth. "Their lips are still, their eyes averted, their faces haggard" is how one storyteller described them.

"Is your mother at home?" Epée is said to have asked, regarding the girls with his usual penetrating gaze.

The young women did not respond or even look up from their work.

Perplexed, the priest sat down nearby to await the mother. Perhaps they have been taught not to speak to men, he thought to himself.

Eventually, the girls' widowed and weary mother returned.

"My daughters are deaf," she explained.

Ah! thought Epée. The extent of their plight became clear. In the eighteenth century, to be deaf was to be virtually alone. If you could not hear, you could not speak. If you could not speak, it was assumed you could not learn. The girls did not attend school; they had few friends and no community. Communication with their hearing mother was sparse, limited to gestures. Their futures were bleak; neither real work nor marriage was likely. As one historian put it, deaf children would probably remain a heavy burden to their parents and "endure an idle and uniform existence."

All that was bad enough, but what distressed their mother most was that without religious instruction her daughters would never be able to take communion. For a time, a kindly neighborhood priest had tried to help by visiting occasionally and showing the girls carvings of the saints, but he had recently died. Now she feared for the souls of her daughters.

As he contemplated the two girls, the abbé knew what he must do. "Believing these two children would live and die in ignorance of their religion if I did not attempt some means of instructing them," he later wrote, "I told . . . the mother she might send them daily to my house."

Resolved to teach the deaf and "to reach heaven by trying at least to lead others there," Epée had then to consider how to teach the girls. For familiar items, he could show them pictures together with the printed words in French (*pain* for a loaf of bread), but what about abstract words like "God" and "duty"? For these, he reasoned that if he had been taught to understand Latin in his native language of French, so should the deaf be taught in theirs. He had seen the way they gestured to each other and communicated among themselves. So, the story goes, Epée turned to sign language and thus began the education of the deaf.

My stack of books on deaf history grew, and in every book I read, the story of Epée and the twin sisters appeared. The details varied and were occasionally embellished. Sometimes it was day, sometimes night; sometimes not just dark but stormy. Sometimes the events took place in the French countryside. Often, Epée is described as "inventing" sign language, though that is an exaggeration. Whether the meeting took place on a dark night, that dramatic bit of storytelling symbolizes the significance of the event: movement from darkness into light. The story is really a folktale of the origin of a culture, according to Carol Padden and Tom Humphries, who together have written several books on Deaf history and culture. "It has come to symbolize, in its retelling through the centuries, the transition from a world in which deaf people live alone or in small isolated communities to a world in which they have a rich community and language."

Since the time of the ancient Greeks, when feats of memory and oratory represented the height of intellectual achievement, the deaf had been considered ineducable. Aristotle believed that they were incapable of learning and of reasoned thinking. If you could not use your voice, he argued, you could not develop cognitive abilities. Beyond an inability to hear or speak, the "deaf and dumb" or "deaf-mutes" as they were then called, were often thought to have a third problem: mental retardation. In nearly every language, the word "dumb" connotes lack of intelligence. And who could prove otherwise? Imagine the frustration of all those trapped minds. It was this barrier to communication that led Samuel Johnson to call deafness "one of the most desperate of human calamities."

At home, with family and those close to them, most deaf people used natural gestures, "home signs," to communicate basic needs and wants. If they were fortunate enough to have other deaf people nearby, they sometimes had a wider repertoire of gestures, but they were effectively barred from the rest of society. Helen Keller famously said that being blind cut you off from things, but being deaf cut you off from people. For her, deafness was the greater affliction. The wild and uncouth behavior of those who couldn't communicate could be indistinguishable from that of the mentally ill. Well into the twentieth century, the deaf were still sometimes put in mental health institutions.

The first to attempt to teach a deaf person any kind of language was not actually Epée but Pedro Ponce de León, a Spanish Benedictine monk born about 1520. In Spain's aristocratic families, there was a higher than average incidence of deafness, most likely the result of intermarrying. In several cases, considerable estates were at risk, because the law of the time prevented the "deaf and dumb" from owning property or writing wills. But if a deaf person could be taught to speak, the law could be nullified. Lack of speech rather than hearing was the decisive factor. Such families had a lot at stake. For his part, Ponce de León didn't want to save the fortunes of his students; he wanted to save their souls. If you couldn't make confession, you couldn't be saved.

And so the deaf had to be taught to speak. Extensive details of Ponce de León's methods haven't survived, but it appears he first taught pupils to associate written words with objects and ideas, and then moved to articulation of those words. He used some gestures and developed a manual alphabet. The records of his successes do live on. One of his most famous students, Pedro de Velasco, left this account:

> When I was a child I knew nothing, like a stone; but I commenced to learn by first of all writing down the things my master taught me. . . . Next, by the aid of God, I began to spell, and afterwards to pronounce with all the force I could, although much saliva came from me. After this I began to read histories, so that in ten years I had read the histories of the whole of the world; and then I learnt Latin.

Teaching the deaf to speak, for Ponce de León and the families with which he worked, was the whole point of the enterprise. It served their legal, economic, and religious purposes. That he succeeded at all was "a breakthrough that shattered for ever the old assumptions," wrote David Wright, a British poet, in his 1969 autobiography, *Deafness: An Autobiography*, one of the first books to examine the history of deaf education.

A few decades later, another Spaniard, Juan Pablo Bonet, worked for the same aristocratic family as Ponce de León: the Velascos. Bonet seems to have learned some of Ponce de León's techniques from them and was the first to publish a book on educating the deaf. In it, he included a chart of handshapes for individual letters—the manual alphabet thought to have been created by Ponce de León. Many are the same shapes used in French and American sign language today.

Epée had a copy of Bonet's book as well as a German book on language and articulation. He considered signs and gestures the native language of those born deaf. Though he thought sign language lacked grammar, he saw in it the "shortest and easiest method" to reach the deaf. "What we cannot cause to enter by the main door," he said, "we can send in through the window." He learned and then codified sign language by devising rules about combinations of signs. That resulted in a manual version of French called Methodical Signs, which followed French rules of grammar and included signs for tense, prefixes, suffixes, agreement, and negation. Epée's system was unnecessarily complicated and actually limited what his students could understand, but that didn't diminish his standing. In the view of psychologist Harlan Lane, who wrote a groundbreaking, if highly opinionated history of the deaf, *When the Mind Hears*, Epée's willingness to ask the deaf to teach him signs and his recognition that he could use them as a vehicle to educate was an "act of humility that gained him everlasting glory."

What's more, Epée prevailed upon the French state to bear the cost of educating deaf students. After his death, the school he founded in Paris in the 1760s became the National Institute for Deaf-Mutes, a haven where students were housed, clothed, and fed, and where they were fully educated in sign language, although some had articulation lessons as well. The school became a model; teachers came from all over Europe to be trained in the "silent" system. After Epée's death in 1789, his successor, the Abbé Roch-Ambroise Cucurron Sicard, published *Théorie des signes*, a grammar and dictionary of sign language. In Epée's lifetime, twelve more similar schools for the deaf were created. Under Sicard, the number rose to sixty.

Sicard's most famous pupil was Jean Massieu. One of six deaf siblings, Massieu began life as a shepherd in a village outside Bordeaux. After coming to the National Institute along with Sicard, who had been running a school for the deaf in Bordeaux, Massieu became a teacher himself, the first deaf man to do so. To fund the school and to convince the world of the capability of sign language to express abstract thought, Sicard and Massieu put on public demonstrations on the third Monday of every month. Sicard interpreted questions from the audience. Massieu answered in sign language, often writing the answers in French on a chalkboard to make clear that they were his own.

"What is hope?" a member of the British parliament asked at one such event.

"Hope is the blossom of happiness," Massieu answered.

"What is time?"

"A line that has two ends, a path that begins in the cradle and ends in the tomb."

"What is intelligence?" Sicard asked.

"It is the power of the mind to move in the straight line of truth," Massieu wrote, "to distinguish the right from the wrong, the necessary from the superfluous, to see clearly and precisely. It is the force, courage, and vigor of the mind."

In other parts of Europe, a different approach to educating the deaf was taking hold. Beginning with Johann Conrad Amman, a Swiss doctor living in Holland, and John Wallis in England, a group of teachers emerged for whom teaching the deaf to speak was the first priority. Their reason was not so much legal, as it had been for the Spanish aristocracy, but religious; they thought it was God's will that man should speak. "The breath of life resides in the voice, transmitting enlightenment through it," wrote Amman. "The voice is a living emanation of that spirit that God breathed into man when he created him a living soul." Amman worked with just a handful of students, all of them the children of wealthy families, and he did teach them to talk, though it sometimes took years. In Britain, Wallis focused on teaching written words.

In Germany, a man named Samuel Heinicke, a former army officer and a contemporary of Epée, worked as a teacher and took on a deaf student around 1754. Using Amman's book as his guide, he opened the first German public school for the deaf. Heinicke believed the deaf had to master speech in order to participate in society. The first "pure oralist," he rejected any type of sign or gesture, even the manual alphabet used by Ponce de León and Bonet, on the grounds that sign language hindered the acquisition of speech. He limited his teaching to lipreading and articulation, forgoing a broader education, and had several impressive successes. Heinicke considered his school a family business and wanted to protect his son's livelihood, so he was secretive about his methods. One, though, was to ask students to feel the vibrations in his throat as he spoke, a common practice. Heinicke's will revealed another, more unusual, technique: taste. He coated his students' tongues with different tastes to "fix" vowel sounds: pure water for "ie," sugar water for "o," olive oil for "ou," absinthe for "e," and vinegar for "a."

That strategy probably died with Heinicke. His underlying philosophy, however, is still appreciated in oral deaf circles. In the history section of AG Bell's website (the site has since been changed), I found

Heinicke listed as the man who "developed the foundations of modern oral deaf education. He believed that language was essential to the process of thinking, and felt that it was critical for children who were deaf or hard of hearing to learn to use spoken language in order to have access to the wider world." He was also, they noted, the "first advocate" of what we know today as mainstreaming.

From the start, the two systems, oral and manual, were in opposition. In 1782, Heinicke sent Epée an extensive argument in favor of oralism. Epée returned fire and the battle was launched. Their disagreement was submitted to the Zurich Academy, which found in favor of Epée, largely, says David Wright, because Heinicke wouldn't reveal his methods. Of course, that didn't settle the matter. For generations to follow, the debate simmered and flared as educational philosophies, scientific thinking, and deaf identities shifted and evolved.

Why so inimical? Any language that fostered communication had to be an improvement over isolation. One would think that the few who were willing to try teaching the deaf, who believed in those early centuries that they *could* be taught, would find common cause. The hostility seemed to spring from a combination of evangelism, egotism, and economics (nearly everyone involved had a vested interest). If you clear away the religious fervor and personal stakes, though, the basic arguments for and against sign and speech established in the eighteenth century were potent enough to persist for another two hundred years.

Learning to communicate through sign was faster and easier than learning to speak. It was, therefore, more widely accessible, putting education within reach of all deaf people, no matter their level of hearing or family circumstances. Epée believed in the importance of helping the deaf as a class rather than working with a few high achievers. By formalizing sign language and putting it in the classroom, he didn't just create a new form of deaf education, he also sparked centuries of questions about the signs themselves. Could they really be considered a language? Were they inferior to spoken language? (Epée himself thought they were.) What were the limitations, if any, to signing? Jean Massieu's vigorous mind notwithstanding, most hearing people—and plenty of deaf people, too—considered sign language primitive, concrete, and pictorial. Those fallacies wouldn't disappear until the 1960s and 1970s, when linguists began studying the language formally. In the eighteenth century as in the twenty-first, there was also the problem of numbers. Was sign language worth pursuing if it meant one could only communicate with other signers? By definition, it seemed, the deaf would be limiting themselves to conversing with one another.

Learning to speak was a more difficult process, sometimes even impossible for those who had been born profoundly deaf. To learn to speak, you must not just hear those around you, you also need to hear yourself. Imagine trying to learn Japanese through a soundproof window, suggests one audiologist. Those in the signing camp were always suspicious of oral successes, wanting to know when and how they lost their hearing and whether any remained. Manualists also argued that the sheer effort involved in learning to talk crowded out educational content. On the other hand, some, especially those with usable residual hearing, were quite capable of speaking. If they could also read lips, properly referred to as speechreading and a skill that is not easily mastered, they could sometimes manage surprisingly well. An oral education could provide access to the hearing world, pushing deaf people not inward but outward.

When a young American carried Epée's sign language to the United States in 1816, the argument over its merits caught a ride on the same ship. A few years earlier, Thomas Hopkins Gallaudet had visited his family's Connecticut home during a brief vacation from his studies at Andover Theological Seminary. Studious, modest, a little frail, and deeply religious, Gallaudet stood in the garden under an elm tree one afternoon and watched his youngest brothers playing a game of fox and hounds with the children of Mason Fitch Cogswell, a prominent local physician who lived next door. Gallaudet paid particular attention to eight-year-old Alice Cogswell, who had lost her hearing completely after contracting "spotted fever" (cerebrospinal meningitis) as a toddler. Despite the Cogswells' best efforts, Alice was not thriving. She had lost what language she'd had, and her family knew that she understood little of what went on around her.

Gallaudet's concern for Alice was "immediate and deep," one of Alice's relatives later wrote. "He at once attempted to converse with and instruct her." Taking the hat off his head, he placed it on the ground and wrote the letters H-A-T in the dirt. Then he pointed from the object to the word repeatedly. When she seemed to understand, he erased the letters and wrote them again farther down the garden. Alice picked up the hat and placed it by the new letters. Gallaudet was thrilled with his success and it marked the beginning, wrote Harlan Lane, of a "consuming interest in deafness." After finishing seminary, Gallaudet spent much of the following year working with Alice.

Though schools for the deaf—both silent and oral—had been springing up across Europe, deaf education in America was nonexistent. A few wealthy families sent their deaf children to Europe, but that option wasn't open to many. Nor was it always desirable. Preferring to keep Alice at home, Mason Cogswell gathered a group of Hartford businessmen to discuss establishing a deaf school. The men resolved to send someone to study European methods and didn't have to look far to find the man for the job. Thomas Gallaudet still lived next door.

Gallaudet set off for Britain in the summer of 1815 intending to combine the best of the oral and manual systems. In London and Edinburgh, he sought out the leading deaf educators, the Braidwood family and their protégés, but they continually put him off and finally suggested an apprenticeship with unacceptably severe restrictions. Meanwhile, the Abbé Sicard arrived in Britain with Massieu and another of his most successful students-turned-teachers, Laurent Clerc. Gallaudet attended one of their fund-raising demonstrations, and Sicard invited him to Paris.

Frustrated by his lack of progress in Britain, Gallaudet eventually took Sicard up on his offer. For a puritanical New Englander, Paris was somewhat horrifying. Gallaudet found the French frivolous and lacking in religious fervor. ("Oh! how this poor heathen people want the Bible and the Sabbath!" he wrote to Mason Cogswell.) But at the National Institute for Deaf-Mutes, he was welcomed with open arms. Massieu and Clerc taught him sign language. He observed any class he wanted. By the summer, Gallaudet had been converted to the idea of using sign language to educate the deaf. He asked Laurent Clerc to return with him to Connecticut.

The American Asylum, the first school for the deaf in America, opened in Hartford in 1817 with Gallaudet as principal and Clerc as head teacher. Like Epée's Institute, the Asylum, which today is called the American School for the Deaf, was a residential school. Clerc and Gallaudet sowed the seeds of American Sign Language by blending Epée's sign language with the many signs already being used by the American deaf. Other schools modeling the Hartford approach soon opened elsewhere. By 1868, there were twenty-seven in the United States, and the number continued to grow. As the name Asylum implies, the schools were designed with the Victorian ethos of public charity in mind. Children lived there from very young ages, and were cared for, but were also kept away from the general public.

Gallaudet's son, Edward Miner Gallaudet, later became the first superintendent of the Columbia Institution for the Instruction of the Deaf and Dumb and the Blind, which opened as a school in 1857 in Washington, DC, on the property of a businessman named Amos Kendall. By 1864, Edward Gallaudet and his supporters prevailed upon Congress to authorize funding for the first National Deaf-Mute College. The Columbia Institution thus became the first center of higher education for the deaf. Today, it is Gallaudet University, still located in the same spot in northeast Washington. It remains the only higher education institution for the deaf in the world that has the right to confer degrees.

Not every school followed the Asylum's lead, however. As in Europe, there was a group in America who preferred that deaf children be taught to speak and read lips. Chief among them was Gardiner Greene Hubbard, a patent lawyer from Boston who, like Mason Cogswell, had a deaf daughter. Mabel Hubbard contracted scarlet fever at the age of five. *Mamma, why don't you talk to me?* Mabel remembered thinking when the illness finally passed. Firmly believing his daughter could speak and learn just like other children, Hubbard hired a teacher to work on her speech and language.

Like Cogswell before him, Hubbard rounded up leading citizens for support, including philanthropist John Clarke, who provided a grant of $50,000 (about $750,000 today). They lobbied the Massachusetts legislature for funding and Mabel, aged nine and an excellent speech-reader, testified at the hearings as an example of what was achievable. Clarke School received its charter from the state in 1867 and opened in the town of Northampton as the nation's first oral deaf school. It is still there, along with satellite schools in four other locations on the East Coast.

The foundations of deaf education in America were solid, but one significant figure had yet to arrive whose views and technological imaginings would shift the course of history and have inconceivable ramifications for the deaf.

Born in Scotland in 1847, Alexander Graham Bell grew up in a household focused on sound and its absence. His grandfather and father were both "elocutionists"; today they would be called speech pathologists. His mother, Eliza, was deaf. Of the three boys, Alec, as he was called (he originally spelled it Aleck but dropped the "k" in America), was the most attuned to his mother and communicated with her by putting his mouth close to her forehead and speaking in a deep voice. Around the dinner table, Alec used the manual alphabet to fingerspell for Eliza and keep her abreast of the conversation. His own ear was said to be "unusually discriminating." At night, lying in bed, he could identify each Edinburgh church bell as it tolled and knew which neighbor's dog was barking.

Melville Bell, Alec's father, spent years creating a phonetic system called Visible Speech, a complex series of symbols depicting each possible human speech sound. It was designed to allow anyone to produce a sound whether they heard it or understood it. His sons regularly demonstrated. In one instance in 1864, the young Bell brothers waited down the hall while Melville wrote out a symbolic version of an obscure sound from Sanskrit along with words from Persian, Hindi, and Urdu dictated by the speech experts in the audience. The boys wowed the crowd by re-creating the sounds exactly. Visible Speech's possibilities for teaching the deaf led both Melville and Alec to begin working with deaf students in the 1860s. Ultimately, though, the system proved too cumbersome and complicated to be workable.

Alexander Graham Bell wasn't as enamored of Visible Speech as his father, yet after he settled in Boston in 1871, he came to see the education of the deaf as his true calling. Although Bell was naturally drawn toward teaching the deaf to speak, his work with one of his students, a boy named George Sanders, was reminiscent of Gallaudet's early lessons with Alice Cogswell. Bell labeled every toy in the boy's room and wrote the name of each object on a card. "George would appear and make his sign for doll," wrote Bell. "He folded his arms and beat his shoulders rapidly with his hands. The doll would be produced and his attention directed to the word 'doll' posted on its forehead. We then compared this word with words written on cards to see who could first find the card with the word 'doll' upon it." Eventually, Bell pretended not to know which toy George wanted until the boy used the proper card to ask for it.

Bell had another passion as well. After spending his days teaching deaf students, he occupied his nights experimenting with ways of using electricity to transmit the human voice. He was an amateur scientist, untrained in the physics of sound and ignorant of the principles of electricity, but nevertheless fascinated and feverishly obsessed with the possibilities. Early on, he used tuning forks to explore the elements of vowel sounds and devised his own instruments to measure the volumes of air in speech. On the ship to North America, he had passed the time poring over Hermann von Helmholtz's 1863 book, *On the Sensations of Tone*, an influential examination of the physics of perception. Bell became convinced that undulating current could mimic the subtle changes in intensity, amplitude, and frequency that make up speech. In 1876, he proved it at his workshop when his assistant, John Watson, famously heard him say over the wire they had rigged between rooms: "Mr. Watson, come here. I want to see you." The telephone Bell designed had a microphone in the mouthpiece that vibrated when you spoke into it and caused a magnet inside a coil of wire to vibrate, too, which generated an electrical current down the wire. The process was essentially reversed at the other end, where the electricity was converted back into sound.

According to his biographer Robert Bruce, Bell was a paradox. "He came to his miracle of sound transmission in working to help those who would be totally unable to avail themselves of it." But there is common ground. "For all the seeming disparity of his interests," wrote Bruce, "there was a basic unity in their tendency: that of furthering communication and human togetherness."

In 1872, at Edward Gallaudet's invitation, Bell visited the American Asylum. While at Hartford, Bell learned some sign language and wrote that he saw its potential as a teaching tool, but his emphasis remained on speech because he believed that allowing the deaf to be part of the wider world was paramount.

His wife shared that view. Mabel Hubbard, daughter of Gardiner Greene Hubbard, began studying speech with Bell in 1873, when she was still a teenager, and married him soon after. She spoke well enough to "pass" in some instances as hearing. "Only the intensity with which she watched her companion's mouth and her distorted vowels when she herself spoke gave her away," wrote Charlotte Gray, whose biography of Bell, *Reluctant Genius*, devotes considerable attention to Mabel. "For Alec, she was living proof that if a deaf person was completely integrated into speaking society, she need not be regarded as 'abnormal.'"

With a deaf mother and a deaf wife, Bell knew what it was to experience the world without hearing. Ironically, these women in his life had a fairly low opinion of the deaf. His mother initially objected to his marriage to Mabel out of worry for their children (she didn't know then that Mabel had lost her hearing to illness). Bell was deeply offended and didn't communicate with his mother for months. For her part, Mabel actively encouraged her husband to give up his work with the deaf, though he refused. "When I was young," Mabel wrote, "and struggling for a foot-hold in the society of my natural equals, I could not be nice to other deaf people. It was a case of self-preservation."

After his invention of the telephone brought money and fame, Bell used his position as a platform and became America's most visible supporter of oral deaf education, founding a research organization on deaf education called the Volta Bureau, which lives on in Washington today as part of the AG Bell Association.

To Deaf historians, Bell committed some unforgivable sins. Worse than his push for oralism were his views on genetics, a hot topic in late-nineteenth-century scientific circles. At the lodge he and

Mabel built as a second home in Nova Scotia, he conducted years of sheep breeding experiments. When he turned his interest in heredity to the deaf, he launched extensive studies of deaf ancestry in places with higher than normal incidence of deafness, like Martha's Vineyard. (At the time, roughly half of the cases of hearing loss in the United States were due to infectious disease.) In 1883, Bell published a paper warning of the risk that deaf intermarriage would result in deaf children. He urged the deaf to socialize with and marry hearing people and even raised the possibility of a law forbidding marriage among the deaf, though he ended by dismissing the idea as unworkable. Even in the context of the age, the paper was deeply offensive to deaf people. There was a storm of attention that sullied Bell's reputation.

Hero to some, villain to others, Bell may have been neither, according to deaf education expert Marc Marschark, who maintains, "He was not as clearly definite in his beliefs about language as is often supposed." Bell's writings hint at the complexities that still reverberate. Bell believed the deaf should strive to join the hearing majority, but he also defended "the de l'Epée language" as "a complete language." He noted that while he preferred an oral approach for the "semi-deaf" or "semi-mute" (i.e., hard of hearing), for the rest he "was not so sure." And Bell regularly set a trap for people by way of demonstrating just how hard speechreading could be.

"It rate ferry aren't hadn't four that reason high knit donned co," Bell would say.

"It rained very hard and for that reason I did not go," a lip-reader would say, diligently repeating what he thought Bell had said.

To the argument that sign language was easier, Bell countered that just because Italian is easier than English doesn't mean Americans should abandon their native tongue.

Until the late twentieth century and the advent of cochlear implants, the most decisive moment in the battle between speech and sign came in 1880 at an international conference of deaf educators in Milan. The conference passed a resolution declaring the "incontestable superiority of speech over signs" and that the simultaneous use of sign "injured" the development of speech and should be prohibited. In short order, many deaf schools around the world switched from teaching sign to "pure oralism" (left to themselves, the students communicated through sign). Between 1900 and 1920, the number of deaf students in America being educated in the oral method went from 40 to 80 percent. Signing was forbidden.

History is never black and white; it is tinted by those who tell it.

According to Harlan Lane, who is hearing and an ardent advocate of Deaf culture, "oralist tradition is a story of greed, plagiarism, secrecy, trickery—but not education. Its aim is speech." Lane's virulence and absolutism put me off, but through him I glimpsed some uncomfortable truths and saw deaf history as many in Deaf culture view it. (Lane has admitted, however, that relatively few deaf people are able to read his books.)

By contrast, David Wright, who learned to talk before he went deaf at seven from meningitis, was a product of a successful oral education, and he ends by making a case against the isolation of deaf schools—at least in his time, the 1930s and 1940s. "The weather of the two worlds, of the deaf and the hearing, is different: in passing from one to the other you have to become acclimatized," he wrote. "From the day I entered the deaf school, I had begun to live a schizoid life, to develop two simultaneous personalities." Wright admits that not everyone would be able to make the choice for the hearing world as he did. "It is almost impossible to exaggerate the narrow scope of the general information of a deaf-born boy whose vocabulary may sometimes be too scanty to allow him to browse over a popular newspaper," he said in describing a class at his deaf school. "If knowledge can be compared to light, most hearing people live in a twilight precinct with one or two brightly lit patches—subjects with which they have special acquaintance. But my companions, it seemed to me, existed in a pitch-blackness shot through with a few concentrated beams of painfully gathered information." Even for Wright, oral education was a steep climb. When he graduated from Oxford in 1942, not more than a half dozen or so other deaf students had graduated from any English university. That fact, he acknowledged, was "a commentary on the recentness and difficulty of the higher education of the deaf."

Seventy years later, the problem can no longer be called recent, but it is still stubbornly difficult.

6

"MARVELOUS MECHANISM"

Into the twentieth century, doctors tried a variety of measures, including the cruel and the crackpot, to repair damaged ears. A French physician, Jean Marc Gaspard Itard, used leeches, pierced eardrums, catheterized the ear, and fractured the skull by striking just behind the ear with a hammer. And Itard was working for Epée. "Medicine does not work on the dead," he concluded, "and as far as I am concerned the ear is dead in the deaf-mute. There is nothing for science to do about it." You can almost hear the harrumph of a man who has been defeated by a problem beyond his abilities. You can also hear the sound of a deaf person's skull cracking. Nineteenth-century doctors, if you can call them that, were operating in the dark. They knew very little about how hearing really worked. The aid they offered was limited to ear trumpets, which, like an expanded version of a cupped hand behind the ear, could be aimed at the source of sound to amplify it ever so slightly.

People knew very little about sound as well. The ancient Greek mathematician Pythagoras conducted experiments with vibrating strings in 500 BC, and Leonardo da Vinci was the first to recognize that sound traveled in waves. But grasping what happened when those waves washed over the ear, how they traveled to the brain and were understood as speech or music or noise, was a problem of another order.

German physician Hermann von Helmholtz was a man of many interests, including physiology, mathematics, thermodynamics, optics, and acoustics. His invention of the ophthalmoscope made it possible to study the interior of the eye, and he advanced the understanding of the perception of color. His 1863 book, the very one that captivated Alexander Graham Bell, has been described as "monumental" and is still used in psychoacoustics, the study of human perception of sound. Helmholtz invented a resonator, a device to intensify and enrich sound by adding vibration, and used it to identify the frequencies of complex sounds. His resonance—or harp—theory described what might happen in the ear when sensing a sound. He captured some of the basics correctly: that air passing through the outer ear set first the middle ear bones and then the fluid of the inner ear in motion. He also theorized that parts of the basilar membrane vibrated "sympathetically" with specific tones and less strongly for other tones, and sent related messages to the nerves.

The capacity to confirm that Helmholtz was onto something and to measure what was really happening in the ear had to wait for Alexander Graham Bell's telephone. It's true that the advent of the telephone made it much more difficult for anyone who didn't hear well to participate in the larger society. For the deaf and hard of hearing, there was irony in the romantic vision many held of the telephone: "With one broad sweep the barriers of time and space are gone and all the world becomes our vocal neighborhood," wrote Harold Arnold, the director of research at Bell Laboratories in the early part of the twentieth century. Yet the effort to perfect the telephone and extend its reach had some unexpected benefits. Western Electric's engineering department was reorganized and rechristened Bell Telephone Laboratories in 1925. Later consolidated with AT&T's engineering department, it would become synonymous with scientific innovation. For the first half of the twentieth century, the Bell Labs building on West Street near the Hudson River in lower Manhattan drew some of the brightest scientific minds in the country and paid them to develop one breakthrough after another, including the vacuum tube, the transistor, and the concept of information theory. Those scientists also, for several decades, knew more than almost anyone else in the world about sound, speech, and hearing.

That research was led by a man named Harvey Fletcher. A Utah native who joined Bell Labs during World War I, Fletcher had done his doctoral work in physics at the University of Chicago with Robert Millikan and taken part in the famous "oil-drop experiment" to establish the charge of all electrons in the universe. At Bell, working for Harold Arnold, whom he had known at Chicago, Fletcher "got into acoustics" as he put it in a 1963 interview. Others warned him that "all there was to know about acoustics had already been discovered," but he proved just how little those scientists knew about how much they didn't know. Fletcher set out to "accurately describe every part of the system from the voice through the telephone instruments to and including the ear," he wrote.

Reading his landmark 1929 book, *Speech and Hearing*, I was surprised by how many of its observations still spoke to my own experience and search for knowledge. I was reminded how much I had taken hearing for granted. "The atmosphere of sounds in which we live ministers so constantly to our knowledge and enjoyment of our surroundings that through long familiarity we have come to feel, if not contempt, at least indifference toward the marvelous mechanism through which it works," wrote Harold Arnold in the introduction to *Speech and Hearing*. "Hearing, we are inclined to consider as little a matter for concern as breathing; and so long as our own faculty remains unimpaired we feel little curiosity concerning the provisions of nature either for ourselves or for others." In a later edition,

Fletcher wrote: "The processes of speaking and hearing are very intimately related, so much so that I have often said that we speak with our ears. We can listen without speaking but cannot speak without listening."

Fletcher and Arnold's research team approached the problem methodically. "The attack was first launched most vigorously on the constitution of speech," wrote Fletcher. If they could establish a reasonable description of average speech, he thought, they could find out what small imperfections and variations affected intelligibility. Their primary weapon was a machine that could better capture what speech actually looked like by creating pictures of waveforms that would be "readily interpreted by the eye." The "high-quality oscillograph" they invented used a telephone transmitter to convert speech waves to electrical waves, which were then magnified with an amplifier and sent into an oscillograph, where they caused a tiny ribbon to vibrate. That motion was photographed on a moving film. Fletcher notes almost offhandedly that in order to create "the perfection of this instrument," they first had to invent three other critical devices: a condenser transmitter that could be calibrated, a vacuum tube to produce the amplification and electrical oscillations, and the basic oscillograph itself.

Like children with a new camera, the Bell researchers used their new device to make wave pictures of a panoply of letters and words. A host of speakers, known to history as "M.A.—Male, Low-Pitched," and "F.D.—Female, High-Pitched," and the like, put their lips about three inches from the transmitter and intoned lists of vowel sounds—the "u" of "put," for example—and words such as "seems" and "poor." The results established some general characteristics of speech sounds. That the pitch of the voice varies with individuals, for instance. A "deep-voiced man" spoke vowels at about ninety cycles per second—or ninety hertz—and a "high shrill-voiced woman" (F.D. perhaps?) at about three hundred cycles per second. They also noticed that when that same man and woman spoke the "ah" sound in "father," the wave pictures looked quite different, "yet the ear will identify them as the vowel 'a' more than 99 percent of the time." Whereas two low-pitched male voices pronouncing "i" as in "tip" and "o" as in "ton" create much more similar pictures, "yet they are never confused by the ear." The Bell team had tumbled to the fact that speech sounds carried some other important characteristic that didn't show up in the waveform. Later, that characteristic was given a name: timbre.

Even I, untrained in reading waveforms and spectrograms, could see that the separate sounds in a simple word like "farmers," casually uttered in an instant, carried detailed and identifiable information that distinguished it from "alters" almost like the fingerprints that distinguish my right hand from my husband's. The very high frequencies in the "f" and the "s" sounds at the beginning and end of "farmers" are so rapid they look like a nearly straight line. The "a," "r," and "m" sounds in the middle show up as peaks and valleys of varying sharpness and depth—the "r" spiking then rolling, spiking then rolling, and the "m" flatter but still undulating like a line of mesas in the desert. All three sounds hovered at the same frequency of 120 cycles per second. The "er" toward the end of the word brought a slight rise in pitch to 130 cycles. "Farmers," I said out loud. Sure enough, I raised my voice in the second syllable, a fact I had never noticed before. From this work, I could draw a direct line to the audiogram chart Jessica O'Gara had given me, so I could see where the main frequencies of various phonemes in the English alphabet fell.

Fletcher and his team spent particular time on vowels. Vowels are distinguished from consonants in the way they are formed in our vocal tracts. Critically, they are also at the heart of each syllable. Syllables, I was going to learn, are an essential ingredient in the recipe that allows us to hear and process spoken language. No wonder all languages require vowels. Expanding on Helmholtz's and Bell's investigations into the complexities of vowel sounds, the men of Bell Labs identified not just the fundamental frequencies of "ah" and "oo," for instance, but also the accompanying harmonic frequencies that readily distinguish one sound from the other. From that, they generated tables showing two primary frequencies—one lower, one higher—for each vowel sound. For telephone engineers, such information "makes it possible to see quickly which frequencies must be transmitted by the systems to completely carry all the characteristics of speech." After World War II, two more Bell researchers were able to use another pioneering device, the spectrograph, to create definitive specifications of vowel frequencies, known as formants. What none of those early researchers could possibly guess was that decades later, a different generation of engineers would use the formant information compiled at Bell Labs to figure out how to transmit the necessary frequencies through a cochlear implant to make speech intelligible to the deaf.

They did see the potential to help in other ways immediately. With his new arsenal of oscillators, amplifiers, and attenuators, Harvey Fletcher could for the first time accurately measure hearing, because he could now produce a known frequency and intensity of tone. He patented the audiometer that was the forerunner of the machines used in Alex's hearing test. His group also created the decibel to measure the intensity of sound as perceived by humans, and they established the range of normal hearing as 20 to 20,000 Hz. The range of speech from a whisper to a yell proved to span about sixty decibels. The new audiometer could measure noise as well, which allowed the editors of the August 1926 issue of *Popular Science Monthly* to note that a Bell Labs device had identified the corner of Thirty-Fourth Street and Sixth Avenue as the noisiest place in New York City. A decade later, Bell scientists capitalized on demonstrations at the 1939 World's Fair and measured the hearing of enough curious fairgoers that they were able to show just how much hearing degrades from the teenage years into late middle age.

Fletcher's newfound abilities and equipment brought him some interesting visitors. When American industrialist and philanthropist Alfred I. duPont couldn't hear what was being said at his own board meetings, he turned to Bell Labs for help. After a childhood swimming accident in the Brandywine River, duPont's hearing had gotten progressively worse, and he was almost entirely deaf as an adult. DuPont told Fletcher that his ability to hear fluctuated. It improved after X-ray beam treatment from a doctor he was seeing, then worsened again. Skeptical, Fletcher asked to accompany duPont on his next visit to the doctor. Beforehand, Fletcher measured duPont's hearing himself and created a picture of his considerable hearing loss using what he called an "embryo audiometer." According to Fletcher, the doctor treating duPont had a very different technique.

> There was a path along the floor . . . about 20 feet long. Mr. Dupont was asked to stand at one end of this. The doctor stood at the other end and said in a very weak voice, "Can you hear now?" Mr. Dupont shook his head. [The doctor] kept coming closer and asking the same question in the same weak voice until he came to about two feet from his ear, where [Mr. Dupont] said he could hear. His hearing level was found to be two feet.
>
> Mr. Dupont then was asked to stand four or five feet in front of an X-ray tube with his ear facing the tube. The X-ray was turned on two or three times. He then turned his other ear toward the tube and had a similar treatment. He then stood in the 20 foot path and another hearing test was made. But this time as he started to walk toward Mr. Dupont [the doctor] shouted in a very loud voice: "Do you hear me now?" As the doctor reached the 10 or 15 foot mark, Mr. Dupont's eyes twinkled and he said he could hear.
>
> I could hardly keep from laughing. . . .

When they returned to the laboratory, Fletcher measured du-Pont's hearing again and found it unchanged. "After that," noted Fletcher, "Mr. Dupont never paid a visit to this doctor."

Fletcher and duPont then turned to the problem of the board meetings. The invention of the telephone had led to the first electronic hearing aids by making it possible to manipulate attributes of sound like loudness and frequency as well as to measure distortion. (Likewise, the invention of the transistor at Bell Labs in the 1950s would revolutionize hearing aid technology by making the devices smaller and more powerful.) One early electronic hearing aid apparently consisted of a battery attached to a telephone receiver. For duPont to hear all the participants in a meeting, Fletcher set up a system with two microphones in the center of the boardroom table and two telephone receivers (one for each ear) attached to a headband for duPont to wear. Hidden under the table was a desk-size set of amplifiers, transformers, and condensers. By using two receivers instead of one, duPont was able to tell where the speaker was. "And that," said Fletcher, "was the first hearing aid Bell Labs ever made." Later, Fletcher made hearing aids for Thomas Edison as well, though Edison later complained that his hearing aids had revealed to him that speakers at the public events he attended said little of interest.

Fletcher wasn't the only one whose work was inspired by the telephone. In the 1920s, just as the Bell Labs team was investigating the properties of speech and hearing, Hungarian scientist Georg von Békésy began zeroing in on just one component of that chain: the inner ear. After completing his PhD in physics in 1923 at the University of Budapest, Békésy took a job at the Telephone System Laboratory at the Hungarian Post Office, which maintained the country's telephone, telegraph, and radio lines. "After World War I, [it was] the only place in Hungary that had some scientific instruments left and was willing to let me use them," he said later. His job was to determine whether making changes in the telephones themselves or in the cables led to greater improvements in telephone quality. His engineering colleagues wanted to know "which improvements the ear would appreciate." At first, Békésy turned to library books for answers. But he soon realized that while a lot was known about the anatomy of the ear, very little was understood about its physiology, how it actually worked. He began studying the function of the inner ear, and the subject became his life's work.

Békésy wanted to see the cochlea in action, and I do mean "see." He collected an assembly line of temporal bones from cadavers at a nearby hospital and kept them in rotation on his workbench. First, he made models of the cochlea based on his samples, then he began to do experiments with the human cochlea. Using a microscope that he designed himself to send strobes of multicolored light onto the inner ear, he watched the basilar membrane, the cellophane-like ribbon that runs the length of the cochlea, as it responded to sound. The setup he rigged allowed him to see a sound wave ripple from one end of the basilar membrane to the other. He also identified critical properties of the membrane, that it was stiff at one end and more flexible at the other. Although the idea that different places on the basilar membrane responded to different frequencies had already been posited, Békésy was the first to see that response with his own eyes: The displacement of one part of the membrane was greater than the rest, depending on the frequency of the tone. His discovery was called Békésy's traveling wave.

After World War II, not wanting to live in what had become Communist Hungary, Békésy continued his work first at the Karolinska Institute in Sweden and then at Harvard. A loner by nature, he never taught a student or collaborated with anyone. Nevertheless, he was awarded the Nobel Prize in Physiology or Medicine in 1961 for his work on "the physical mechanism of stimulation within the cochlea." Nobel Prize or no, we know today that there are at least two fundamental problems with Békésy's work. One is that his subjects were dead. The auditory system is a living thing and responds more subtly when alive than dead. Secondly, in order to get any response from the cochlea of a cadaver, he had to generate noise that was loud enough (134 dB) to wake the dead, so to speak. As a result, the broad response Békésy saw didn't accurately reflect the finesse of the basilar membrane. Despite its limitations, Békésy's traveling wave represented an important advance. In a recent appreciation of his work, Peter Dallos and Barbara Canlon wrote, "This space-time pattern of vibration of the cochlea's basilar membrane forms the basis of . . . our ability to appreciate the auditory world around us: to process signals, to communicate orally, to listen to music."

Békésy had narrow shoulders, but in the best scientific tradition, many who came later stood upon them. Back at Bell Labs, in the 1950s, later generations of researchers used Békésy's work to build an artificial basilar membrane and then, in the 1970s, to devise computer models of its function, all of which would prove critical in the digital speech processing that lay ahead.

Jean Marc Gaspard Itard had been dead wrong about the possibilities of science.

7

WORD BY WORD

The day after Alex got his hearing aids, he and I went on an errand. As I unstrapped him from his car seat, I discovered he had yanked out both earmolds. At least they were still connected to the safety clip—his had a plastic whale to attach to his shirt and long braided cords like those for sunglasses. But when I tried to reinsert the earmolds, everything looked wrong. Alex had twisted them out of position. I stood there with the tangle of nubby plastic and blue cord in my hand and realized that in spite of the lesson Jessica had given me, I had absolutely no idea how to restore order—which was left, which was right, whether they were backward or forward, or how to begin to put them into his ears.

"Well, shit," I said out loud. "Shit, shit, shit."

Alex smiled. At least I was free to curse in front of him. The word "shit" had high-frequency "sh" and "t" sounds that he would never hear. After the frustration of weeks of uncertainty and disequilibrium . . . here I was, on the sidewalk, lost.

Fortunately, we were about to visit the preschool program at the Auditory/Oral School in Brooklyn, and I'd been given a fresh reminder of the benefits of an environment where people were familiar with what it is to be deaf or hard of hearing. It was just dawning on me that one of my new roles was going to be serving as Alex's IT Help Desk, and I was sorely unprepared.

Piling the tangled equipment onto the top of his stroller, I made my way to the front door.

"Um, we need a little help putting these back in," I confessed when I got inside. "We're new at this."

The director of the school smiled and picked up the hearing aids. "See how the mold curves?" she said, indicating the way the plastic followed the line of Alex's ear canal. Gently, she pulled his earlobe back and popped one aid into position. Then she did the same on the other side. Ten seconds and it was done.

In making choices about education and communication in the deaf and hard-of-hearing world, people talk about "outcomes." Since Alex had usable hearing and our desired outcome was talking and listening, we had decided to look at oral programs. These were the "option" schools my friend Karen had been talking about a few weeks earlier. All would provide explicit language instruction to get Alex beyond "mama," "dada," "hello," and "up."

I had thought Karen was out of her mind to suggest that a two-year-old child might travel an hour from home for school. The additional complication of having to get two other children, then seven and four, to and from school near our house stymied me. But Mark is good at making the impossible seem possible, and far less worried than I am about spending money to solve problems. He immediately threw out solutions: babysitters, car services, an Ecuadorean taxi driver we knew who might be willing to help, and so on. Soon we had a plan. I wasn't willing to put Alex on the school bus yet, so we split the week between me and a pair of babysitters and ultimately chose Clarke, a Manhattan satellite of the school founded by Mabel Hubbard's father, in part because it was reachable by subway. There, Alex would spend every morning bathed in words and language.

• • •

In the same way that an aspiring athlete has to train and strengthen muscles, Alex had to practice learning to talk. Speech production is a motor skill like kicking a ball or picking up a raisin. We don't think of it that way because it doesn't usually make us sweat or even require much effort once we've mastered it, but a babbling baby is training her vocal system to produce the sounds she's been hearing through the first months of life. The attempts are tentative at first. She knows she's getting close when the adults in her world get excited about the noises she makes. The sounds get more and more confident until they come out as words. Alex had missed all of that.

Sound is produced by vibrations and columns of air. Our bodies provide both. In all spoken languages, the fundamental speech sounds are similar because the range of possibilities has physical limits. Words begin in the lungs, which serve both as a store of air and as a source of energy. That air is pushed out of the lungs and, on its way to being transformed into the sounds of speech, it passes along the conveyor belt of our vocal systems.

Lodged in the top and front of the trachea, the larynx is made mostly of cartilage, including the thyroid at the front that forms the Adam's apple. Inside the larynx are the vocal cords. They're sometimes called the vocal folds, and that's a more accurate term, as there is nothing cordlike about the vocal cords. They are pieces of folded ligament that meet to make a V-shaped slit (the glottis) that closes to stop air or opens to let it pass through. To produce the "d" in "idiot," for example, air is stopped entirely. For soft sounds like the "f" of "farm" and the "s" in "sunny," the vocal cords are completely open, and the feathery or hissing sounds can go on indefinitely, which is why those consonants are described as "continuant." And why they had so much less going on in the picture of "farmers" created by the Bell Labs oscillograph. When the vocal cords rapidly open and close, they create a vibration that allows us to make vowels and the sounds of "voiced consonants" such as "v," "z," "b," "d," and "g." If you watch the changing shape of your lips when you make a "p" sound, an "o," or an "f," you can get an idea of the movement of the vocal cords inside your throat, and you can feel the difference between voiced and unvoiced sounds if you make a "zzz" and then a "sss" with a finger resting on your Adam's apple.

To whisper, we keep our vocal cords in the same middle position as for the "h" of "hill." The louder the whisper we want to make, the closer together we bring our vocal cords, so that the word "hill" spoken in a loud whisper results in more air leaving your lips than saying it in a normal voice.

Leaving the lungs and vocal cords as somewhat amorphous buzzes and whooshes, the flow of air is further refined—stopped and restarted, pushed and pulled, narrowed or flattened—by the tuning we do in our mouths when we vary the shape and relative position of the palate, tongue, teeth, and lips. When speech pathologists talk of plosives or fricatives, for instance, they are describing what we have to do in this last stage of the conveyor belt. The plosives ("p," "b," "t," etc.) require us to block the flow of air somewhere along the way, usually in the mouth. The fricatives ("s," "f," "sh," etc.) are made by narrowing the air flow to form turbulence. To form the liquids ("r" and "l") we raise the tip of the tongue and keep the mouth a bit constricted. "M," "n," and "ng" are nasals; "w" and "y" are semivowels. Speech sounds are further identified by place of articulation—labial (lips) or dental (teeth), for example. So a "p" is an unvoiced labial plosive, and a "th" is a voiced dental fricative. Like the Linnaean system of biological classification into kingdom, phylum, class, and so on, this way of organizing elementary features was a breakthrough when it was invented in the 1930s because it captures all of the speech sounds of the world's languages.

All of this matters when you're learning to talk because you have to know how to form the sounds you want to make. Most of us can go our whole lives blissfully ignorant of all the unvoiced labial plosives we produce in a day. But we had to learn, too. As newborns, we had a vocal tract that was not yet capable of the fine motor skills necessary for speech. An infant's larynx is too high in the throat, and the tongue fills the mouth. But the system matures quickly, and usually by four months, babies begin playing around, babbling, with their teeth and tongues. By trial and error, most get there naturally, though some sounds are harder to master than others. Our son Matthew had trouble with "r's" and "l's" until he was six. Words like "girl" and "earth" were almost unintelligible, which wasn't all that uncommon. "R" and "l" tend to be among the last phonetic sounds that children master.

Without hearing aids or a cochlear implant, someone who is profoundly deaf can not only not hear others, he also cannot hear himself. The only way for him to form spoken words is to memorize where exactly to put his tongue, how to form his lips, and what it feels like if he touches his larynx for each separate sound.

For Alex, on the other hand, his new hearing aids were now bringing him information about sound he hadn't had access to before. Relying on residual hearing, hearing aids amplify sound to bring as much as possible into the range necessary for understanding speech. They had come a very long way since Alfred duPont's desk-size boardroom set. Today, they are digital; audiologists can program them very precisely to an individual's hearing loss. It was as if we had pushed the reset button on learning language, but with a lot of ground to make up.

Perched on a wooden stool in the hallway of the Clarke School on a Monday morning in April, just after Alex's second birthday, I watched through an observation window as he worked with a speech language pathologist named Alison for the first time. A headset allowed me to listen as well.

"Choo choo," Alison said as she rolled a wooden engine along a piece of track.

"Push," she said, showing him how she moved the train. "Push."

She blew a bubble. "Pop!"

She held her hand over his and touched his chest. "Me," she said, tapping his chest with his own little hand, "me."

Alex, watchful and shy, blew some bubbles, but he didn't say "pop" or "me." Instead, he pointed to a toy farm on her shelf that had caught his eye. When Alison held the cow to her face and said "moo," Alex didn't imitate her. But he laughed and pointed to the next animal.

Though he enjoyed the animals, Alex was anxious and tearful in his first few weeks at Clarke. Each week, his teachers of the deaf and therapists sent home notes on his progress.

"Limited verbal output today; still a bit fearful during transitions," his teacher noted in the second week.

"Alex cried upon separation from his mother . . . ," the next day's note read. "He repeated 'pat-pat-pat' and 'rooooll' during play with Play-Doh; when he saw his mother through the window/door, he fell to the floor and cried."

Note to self: Stay well back from the window.

The next week was better: He "actively stomped his feet" for "If You're Happy and You Know It" and "enjoyed pasting pictures to the con-struction paper." On the other hand, he "monopolized the glue sticks."

Play never sounded so un-fun. But if the clinical detail effaced the joy, I knew there was a point. For these children, no word could be taken for granted. Everything had to be introduced and repeated. The notes came with lesson plans for every week, listing vocabulary to be targeted and all the themed activities that would be used to teach the new words. As for all children, one of the goals of preschool was to learn how to get along with others. A few of Alex's new class-mates had issues beyond hearing: behavioral problems, or motor skill delays. Glue sticks notwithstanding, I was relieved that there was no sign in Alex of some of the disruptive or asocial behavior I saw in some of the other children.

Rather, Alex seemed to be the kind of kid who needed to learn to stick up for himself. He had to be taught to say "me" or "no" if a class-mate took the truck he was using or tried to steal his Goldfish.

Beyond his behavior, the daily reports teemed with approxima-tions, verbalizations, modeling, mouthing, gesturing, cuing, and so on. It took me a little studying to learn how to decode the notes.

"Alex assumed articulatory posture for /w/ sound without sound emission. . . ." Translation: He mouthed the "wah, wah, wah" of the babies crying in the "Wheels on the Bus" song.

"Alex approximated production of 'cut cut cut' and 'knock knock knock' imitating clinician's productions with accuracy in number of repetitions and syllable length." He said (probably) "cu, cu, cu" and "na, na, na."

By June, he was combining "actions with labels": "wash baby," "come doggie," "go car."

By August, according to a list I kept, he had close to a hundred words, although most were approximations. He called himself "Ala." He could handle Mama, Dada, and Matty, but he had trouble with "j" and "s," so big brother Jake and his babysitters, Jacky and Sean, who took turns tak-ing him to school, all got "d" at the front of their names: Dake, Dacky, and Dawn. Other new words and phrases were painstakingly added:

Eyes, nose, hat, milk . . .

Up, down, shoes, juice . . .

Water, boat, bike, away . . .

Uh-oh, more, all done . . .

Stop it. No touch. I love you.

Because he couldn't hear high frequencies, he left out all "k," "t," and "s" sounds. He said "ow" for "cow" and "um" for "come."

He was still very quiet and observant. Mostly he was compliant, and if he was frustrated, he didn't show it. So it came as a surprise one day when he threw his hearing aids into the street as he rolled along in his stroller. They were nearly run over by a bus.

"I'm relieved to hear it," said one of his teachers when I told her.

"What?!"

"It's developmentally appropriate," she said.

He was two after all.

Alex had an audiogram. He had hearing aids. He was in a specialized school. He was making progress, and the flurry of activity had apparently slowed into routine. One question remained: Why had this happened?

Neither Mark nor I knew of anyone in either of our families who had been deaf or hard of hearing as a child. Nothing dramatic had happened during my pregnancy: no infections that we knew of, no trauma, no worrisome blood tests or ultrasounds. The only drama had come at the end, when Alex arrived four weeks earlier than expected, which doctors classified as preterm but not premature. I couldn't help but wonder if I was responsible somehow, even though the rational part of my brain knew that was unlikely and that such thinking was counterproductive.

"We will probably never know why this happened," Dr. Dolitsky told me. "You could spend a lot of time and money trying to figure it out, but I don't think that's worth it." Many cases of congenital hearing loss, meaning a loss present at birth, are thought to be hereditary, but that didn't seem likely in our case. Nonhereditary causes included infections during pregnancy. For example, an outbreak of rubella (German measles) in the 1960s affected many pregnant women and their fetuses and led to a higher incidence of deaf and hard-of-hearing babies, the so-called Rubella Bulge. Maternal diabetes, prematurity, toxemia, lack of oxygen, or complications with the Rh factor in blood could all cause hearing loss. Or a young child could have acquired hearing loss after birth from ear infections, meningitis, ototoxic drugs (medicine that damaged the auditory system), measles, encephalitis, chicken pox, mumps, flu, or head injury. Babies could even lose hearing from noise exposure, though that is far more common in adults.

There are, however, a few causes of hearing loss that are linked to other medical issues. Those, said our doctor, are things you would want to know about. His recommendation was a medical workup that included only tests that were either easy (like a blood test) or that allowed us to rule out complications. I went home with a stack of prescriptions: EKG, ultrasound, CT scan, and so forth.

First, we looked at genetics, which, now that medicine has succeeded in reducing infectious disease, accounts for about half of all hearing loss in newborns. In recent years, scientists have begun to isolate genes related to hearing loss. In three-quarters of inherited cases, the cause is an "autosomal recessive" gene called connexin 26, which can be passed on if each parent carries it. Alex's blood test was negative for connexin 26. Then we tested his heart, as there is evidence of a correlation between hearing loss and cardiovascular disease. Doctors don't yet know why that should be but hypothesize that the link might be impaired blood flow that damages the sensitive inner ear and can also damage the heart. Alex's EKG was normal. Next, we looked at his kidneys. Since both the ears and the kidneys form in utero around seven weeks, it's possible that if the fetus suffered some sort of trauma that affected the development of the ears, it might also have damaged the kidneys. Alex's kidneys were fine. Nonetheless, for a time, I replayed the seventh week of pregnancy in my head. We had been on a vacation in Italy with my extended family, a trip planned before I'd known I was pregnant. Because I was in Italy and because it was my third pregnancy, I allowed myself a few glasses of wine and some caffeine to battle the jet lag. Was that what did it? I knew better. There would be a far higher incidence of hearing loss in French and Italian babies if a glass of wine could cause it, but I wished I had never left Brooklyn.

The final test was a CT scan, which would give us a look at the inside of Alex's head. It required him to be completely still. Because that was an impossibility for a two-year-old, he had to be sedated again. Out came a big needle, and I held his hand. He whimpered but quickly fell asleep. Once he was asleep, the technicians laid him onto the scanner bed, and I moved to the control room, where I could watch through the window and see the computer screens.

The CT scanner consisted of an adult-size white bed that moved through a circular opening in a big white machine. It looked like a portal on a spaceship. CT stands for "computed tomography." *Tomos* is Greek for "slice," and that's what the machine captures: slices. It uses X-rays to make two-dimensional images of multiple layers of a three-dimensional object—in this case, Alex's skull. Together, those slices would be compiled to create a detailed picture of his anatomy.

There was my small son, dwarfed by the bed, his soft arms and legs peeking out of his orange shorts and blue-and-orange striped shirt, unconscious on the other side of the glass. Suddenly, overwhelmed by his vulnerability, I had to turn away and lean against the wall. Images were collecting on the screen, and the technician began to describe what he was doing.

"Uh-huh," I managed to respond, my voice cracking a little. "Okay."

When it was all done, I cradled Alex in my arms and carried him out of the hospital and into the sunlight of a bright June day, unwilling to let go.

A few days later, the phone rang. It was Dr. Dolitsky. "We found it," he said. The CT scan had revealed that Alex had a congenital deformity of the inner ear similar to something known as Mondini dysplasia or Mondini deformity. It was rare, affecting fewer than 200,000 people in the United States, and it meant that his cochlea had failed to form completely. It seemed unlikely, then, that he had ever had normal hearing, despite the early hearing test that said otherwise.

In addition, he had a second condition that often accompanies Mondini dysplasia: enlarged vestibular aqueduct (EVA), also known as large vestibular aqueduct syndrome (LVAS). Vestibular aqueducts are circular bony canals that look like soda-can pull-top rings and sit just above the inner ear. They help us balance. In Alex and others with EVA, one or more of the vestibular aqueducts is larger than normal, meaning it's more than one millimeter in diameter, roughly the size of the head of a pin. That makes it susceptible to injury. A bump on the head or a change in pressure could result in a rupture in the sac of endolymphatic fluid that is attached to the aqueduct. When that happens, the fluid inside drips down onto the inner ear, with which it is not chemically compatible. The result is further damage to the hair cells of the inner ear. Nearly every child with EVA develops some hearing loss, and according to the National Institute on Deafness and Other Communication Disorders, 5 to 15 percent of children with sensorineural hearing loss have EVA.

"The usual recommendation is no contact sports," Dr. Dolitsky told me over the phone, explaining that strategy as a way of eliminating one category of risk. "And nothing that would involve a big change in pressure." He ran down the list: karate, soccer, football, scuba diving . . . I scribbled notes on the paper that was at hand, a bright blue-and-green pad for making grocery lists that seemed a bit too lighthearted for the occasion. "You will have to make your own decisions," he said. "Unfortunately, in some instances, a drop can be caused by an airplane flight or even a big sneeze."

We finished our conversation and I hung up, a little stunned. I was glad to have an explanation, but now the situation could change again at any moment.

The boys were playing nearby in the living room, and I turned in time to see Alex follow the lead of his big brothers and leap off the couch. All three then rolled around wrestling on the floor. Scuba diving was not going to be my problem.

8

THE HUB

My rental car bumped down a long dirt road flanked by fields of crops. I glanced at my scribbled instructions and kept to the right. In the distance, I finally spotted a tall, thin, elderly man standing in front of a one-story frame house, part of a small group of buildings at the end of the road. He waved as I pulled up in front of him.

"You found me," he said with a smile.

"At last," I answered.

I had been driving for hours to reach this small town in northwestern Oregon between Eugene and Portland, but that wasn't what I meant. Dr. William House had been on my mind for some time.

As a surgeon in Los Angeles in the late 1950s, House had been the first American to seize on the idea of electrically stimulating the ear. The device he ultimately created, which was marketed by 3M, became the first to win FDA approval in 1984. Adoring patients called him Dr. Bill and considered him a hero. Ear surgeons today describe him as a "creative genius." "Without him, we might not have a cochlear implant," says Dr. Marc Eisen, a Connecticut otolaryngologist who has written about the history of cochlear implants.

Yet House has also been roundly criticized over the years. Early on, while he was developing his implant, he was practically shouted down at scientific meetings. Establishment researchers—most of them on the East Coast—thought his idea would never work. "Otology needs a new surgery; this isn't it," said Harold Schuknecht, Harvard Medical School professor and chief of otolaryngology at Massachusetts Eye and Ear Infirmary, at one conference. "If I tell you that a lead balloon will not fly, and you go out and build a lead balloon and it does not fly, what have you learned?" demanded another prestigious scientist. Even if it did work, they didn't like the way House was going about it.

Nonetheless, House built a cochlear implant that worked—his lead balloon flew, for a time anyway. It has since been replaced by more sophisticated devices, a development he resented and resisted. When I brought up House's name with basic scientists rather than doctors, some dismissed him as a kook or a "crazy surgeon." I wondered what the truth of it all was, and I wasn't sure House would want to talk to me.

Then one day my phone rang. The man on the other end was cheery and welcoming even if he sounded every one of his eighty-seven years.

"I got your letter," he said. "I'd be happy to have you visit."

The idea of using electricity to treat deafness would seem to require a futuristic faith in the possibilities of science, but it dates to the late 1700s. Electricity was a relatively new source of fascination then, and curious scientists everywhere were working to understand its principles. The Italian physician and physicist Luigi Galvani kept an electrostatic machine in his laboratory. One day, just as the machine was generating sparks, an assistant happened to touch the sciatic nerve of a dissected frog with a scalpel. The frog's leg muscle twitched. Intrigued, Galvani set up a series of experiments and succeeded in making the frog's muscle twitch under a variety of conditions. He had discovered bioelectricity, the fact that our nerves use electricity to send signals, though he didn't quite understand what he was seeing. The force Galvani called "electric fluid" or "animal electricity" looked to him like an innate, unique form of energy. The 1791 publication of his finding stirred excitement for its potential in treating medical conditions.

Galvani's compatriot Alessandro Volta, a professor of physics at the University of Pavia, was paying close attention. Unlike most of their colleagues, Volta didn't believe in Galvani's "animal electricity" theory. Instead, Volta guessed it was contact between two dissimilar metals touching the frog's leg that caused the stimulation—the frog was simply a conductor. Galvani answered by inducing the same response with two pieces of the same metal. Back and forth the two scientists went, trading competing theories. The disagreement was cordial (it was Volta who coined the term "galvanism" in honor of his friend), but it was persistent and public. Today, we know Galvani was correct in recognizing that the electricity occurred naturally in the animal tissue, and Volta was right that this was not "animal electricity."

In his efforts to develop his own theory, Volta experimented with metals alone. He stacked pairs of silver and zinc disks separated by brine-soaked pads. When he touched the top and bottom of the pile with a wire, an electric current flowed through the pile and along the wire. The contraption became known as the voltaic pile; Volta had invented the battery.

To explore the idea of using electricity medically, Volta applied his voltaic pile to the body. First, he made muscles contract. Then, connecting his battery to the optic nerve, he generated a flash of light when he touched any part of his face. Next he turned to hearing. Into his own ears he inserted two metal rods connected to a circuit of thirty or forty cells with about fifty volts of power. Electricity crackled through him, and he later described the boom he experienced:

"I received a shock in the head and some moments after I began to hear a sound, or rather noise in the ears, which I cannot well define; it was kind of a crackling with shocks, as if some paste or tenacious matter had been boiling."

The scientific world was intrigued yet cautious. Few were willing to repeat the experiment on themselves. The connection between electricity and hearing had been made—literally—but for the next century and a half, there was little progress.

Then in Paris, in the 1950s, electrical hearing became reality when "the impossible" was tried. André Djourno was a neurophysiologist who studied medical applications of electricity at the Institut Prophylactique (today called l'Institut Arthur Vernes). Working with rabbits and guinea pigs, he was stimulating nerves by implanting induction coils. Charles Eyriès was the chief of the hospital's head and neck surgery department and an expert in facial nerve repair. In February 1957, a fifty-seven-year-old patient, Monsieur G., came to Eyriès in fairly bad shape. Surgery to remove two large cholesteatomas, a skin growth that pushes from the middle ear into the inner ear, had left the man deaf in both ears and with extensive facial nerve paralysis. Eyriès wanted to reanimate the facial nerve with a graft. A colleague proposed that perhaps the man's deafness could be addressed as well if Eyriès implanted one of Djourno's induction coils during the surgery. The patient, Eyriès wrote later, had "expressed the desire that the impossible be tried in order to put an end, however imperfect, to his total deafness." Since he would be undergoing surgery anyway, it was thought he had nothing to lose. For his part, Djourno was fascinated by the opportunity.

The facial graft repair (using fetal tissue) was a success, but Eyriès and Djourno found the cochlear nerve "significantly shredded." They put the active electrode into the remaining stump of the nerve and placed the induction coil in the temporalis muscle. During the operation, they tested the device with a variety of stimuli: bursts of low-frequency current at a rate of fifteen to twenty pulses a minute, then low-frequency alternating current, and also words spoken into a microphone. From the start, Monsieur G. heard sounds. He could discriminate the loudness or softness of sounds but not their pitch. Speech was unintelligible. During extensive rehabilitation in the following months, increasingly complex signals were tried. Eventually, the man was able to tell the difference between low frequencies, which sounded to him like "burlap tearing," and high frequencies ("silk ripping"). He could hear some environmental noises and a handful of words, but he never understood speech. Within a few months, the implant stopped working. Eyriès and Djourno found that the electrode in the muscle had broken; a second implant also failed. Eyriès washed his hands of the project. Djourno tried again with a different surgeon and a new patient, but the young woman was less enthusiastic and Djourno's funding ran out.

Eyriès and Djourno reported on their work in several French medical journals. It received little attention in other countries, although at least one researcher suggested to me that they deserve the credit for inventing the cochlear implant. Their successful surgery was, however, mentioned in a short article in an English-language publication that was seen by a patient in California. He clipped out the article and brought it to his otologist, who happened to be Bill House.

"The light went on," House told me as we sat in his small living room in Oregon. A few years earlier, he had moved up from California to live next door to his son, David. "See, we'd known that putting electricity near the ear you get a sound, put it across the eye and you get a flash of light. So the nervous system is very highly attuned to telling you what's happening. [The French report marked] the first time I realized a patient had a total loss of the cochlea and could still hear with electrical stimulation."

. . .

How could it be possible to hear with a nonfunctioning cochlea? The cochlea is the hub, the O'Hare Airport, of normal hearing, where sound arrives, changes form, and travels out again. When acoustic energy is naturally translated into electrical signals, it produces patterns of activity in the thirty thousand fibers of the auditory nerve that the brain ultimately interprets as sound. The more complex the sound, the more complex the pattern of activity. Hearing aids depend on the cochlea. They amplify sound and carry it through the ear to the brain, but only if enough functioning hair cells in the cochlea can transmit the sound to the auditory nerve. Most people with profound deafness have lost that capability. The big idea behind a cochlear implant is to fly direct, to bypass a damaged cochlea and deliver sound—in the form of an electrical signal—to the auditory nerve itself. "The inner ear is a pretty beautiful natural platform for stimulation in the sense that from very early childhood, it's in a stable adult size and form," says auditory neuroscientist Michael Merzenich, who was instrumental in a later stage of cochlear implant development in the 1970s. "Several surgeons got at the idea that conceivably you could excite it and recover enough hearing to be useful."

To do that would be like bolting a makeshift cochlea to the head and somehow extending its reach deep inside. A device that could replicate the work done by the inner ear and create electrical hearing instead of acoustic hearing would require three basic elements: a microphone to collect sound; a package of electronics that could process that sound into electrical signals (a "processor"); and an array of electrodes to conduct the signal to the auditory nerve. How best to build those pieces was anyone's guess. Some of it at least seemed achievable with time. Electronics could be engineered, for instance, and tolerable levels of stimulation for the tissues involved could be determined through animal studies. More difficult was the question of how to excite discrete groups of nerve fibers. Even if those technical problems were solved, and electrodes successfully and safely implanted, a basic science problem remained, to which no one in the 1960s had an answer: what signal to send.

The processor had to encode the sound it received into an electrical message the brain could understand; it had to send instructions, and no one knew what those instructions should say. They could, frankly, have been in Morse code—an idea some researchers considered, since dots and dashes would be straightforward to program and constituted a language people had proven they could learn. By comparison, capturing the nuance and complexity of spoken language in an artificial set of instructions was like leaping straight from the telegraph to the Internet era. It was such a daunting task that realistically, most scientists thought the best they could hope for was to make speechreading easier. "The more a researcher knew about auditory neurophysiology or speech acoustics, the more confident he was that implants could not provide a high (or even useful) level of speech understanding," wrote Michael Dorman and Blake Wilson in an account of some of the early research. The few who "imagined that you could just replace the signals in the ear in some magical way," says Merzenich, didn't really know much about "how the complexities of sounds that would be meaningful, like the sound of oral speech, had to be represented across the nerve to the brain."

No one was sure what, exactly, the brain needed to hear to distinguish between a dog barking and a baby crying, or to know to get out of the way when a car horn blows. They doubted it would ever be possible to make an implant that allowed a child to hear his mother say "I love you." Before sound could fly direct to the auditory nerve, someone would have to reinvent the airplane.

Bill House aimed to try. In his first year of private practice in Los Angeles, House saw two families with two-year-old children they suspected were deaf. At the time, there was no test to uncover hearing loss at earlier ages. House found it painful to tell parents their children were deaf. "I felt I was presenting a very bleak outlook to these parents," he said. "What I had to offer seemed very inadequate." When he learned about the work of Djourno and Eyriès, he immediately saw the potential to do more and resolved to pick up where the French had left off. "I felt if there was anything we could do, we should."

It was an attitude he learned from his father. House grew up on a five-acre ranch in Whittier, California, not far from Los Angeles. Although they kept a few cows and grew avocados, oranges, and lemons, all of which were the responsibility of Bill and his brothers, the family business was dentistry. His father, Milus, set up a private practice in an old barn. Milus House wasn't the kind of father who played ball with his boys or took them fishing, but he made a big impression when he talked about the satisfaction he gained from fixing serious dental problems that affected patients' emotional and physical well-being. "I could feel the joy he had as he talked about 'fixing mankind,'" wrote Bill House years later. "I knew then that I too wanted to be a 'healer.'" After two years as a dentist in the Navy, House went to medical school to specialize in ear, nose, and throat surgery, then narrowed that down to ear surgery, or otology. In 1956, he joined his half-brother Howard, who was ten years older, in practice at what became known as the House Ear Institute, a leading West Coast center for otolaryngology then and now. (Today it's called the House Research Institute and is run by Howard's son and Bill's nephew, John House.)

Bill House told me he had been a mediocre student. Later in life, he realized he probably had dyslexia; writing and reading were always a challenge. But he excelled at working with his hands—a skill his dental training helped hone—and became a top-notch surgeon. He was also driven to solve problems. He spent hours in the morgue, practicing surgical techniques and new approaches on unclaimed bodies. His wife, June, was a registered nurse. If she could get a babysitter, she joined him in the morgue to hand over his instruments.

One of House's first innovations was to introduce and improve upon the use of a surgical microscope for ear surgery. He developed new surgical techniques for acoustic neuromas that he says helped reduce mortality from 40 percent to less than 1 percent and preserved the facial nerve. For patients with Ménière's disease, a disease of the inner ear whose symptoms include vertigo, vomiting, tinnitus, and hearing loss, he created a small device, a shunt, to correct the condition. His most famous patient was astronaut Alan Shepard, who developed Ménière's after his first space flight. With one of House's shunts in place, Shepard went back into space as commander on Apollo 14. A grateful Shepard invited House and his wife to Cape Canaveral for the liftoff.

When House got interested in cochlear implants, he enlisted the help of electrical engineer Jim Doyle. As a first step, Doyle built a battery-operated amplifier and electrode that could be applied to the round window, the membrane that leads to the inner ear. House tried it out during middle ear surgery with three volunteer patients who had lost their hearing after developing speech. Under local anesthesia, he lifted the eardrum and delivered small alternating currents to the round window. All of the patients heard sounds; furthermore, the sounds seemed related to the frequency and intensity of the electrical stimulus. Excited, House had Doyle set to work creating an implantable version. The new device had a silicone-covered coil to generate current, amplifiers, and an electrode, but nothing as fancy as a speech processing program. The coil would be placed in the mastoid bone and the electrode in the cochlea. Wires running from the coil ended in a plug in the skin behind the ear. In the lab, patients would be connected—plugged in—to an electronic stimulator that could send signals through the system to the auditory nerve. When not in use, the plug behind the ear would be covered with a bandage.

Two adult patients volunteered, and they were implanted early in 1961. For two weeks, they underwent testing with a variety of sounds.

"They could hear the sounds, and it was obvious that even though the sounds were not clear, the devices would be of great help for environmental warning sounds and lipreading," says House. Soon, however, redness and swelling appeared around the external wires. "I got kind of scared at that, and I had to take it out," says House. "They were disappointed and so was I." He had run up against the problem of biocompatibility.

Even in the 1960s, when research protocols did not preclude putting an untested device straight into human subjects, many researchers considered House's approach at best unscientific and at worst dangerous. Certainly, it was emblematic of the strengths and weaknesses of House's hands-on style. "He started building things and putting them in patients," says Eisen. "He got his information from talking to people, not reading books. You'd get in big trouble if you did what he did today, but . . . if he'd read more books, he might have believed all the people who told him it couldn't be done."

Biocompatibility was not the only problem. Doyle, the engineer, had made grand claims of eliminating deafness, which caught the imagination of technology buffs and science-fiction fans: "Electronic Firm Restores Hearing with Transistorized System in Ear," read a typical headline in *Space Age News*. Doyle also saw commercial potential and set about creating a company around the device, a move that was at odds with House's more altruistic vision. The ensuing publicity was too much too soon. It offended other doctors and scientists and led to a deluge of calls from people seeking help whom House had to send away. According to House, he told Jim Doyle he was going to put the project on hold until they had biocompatible materials and asked for a full report on the materials and electronics. Doyle refused. "He said I was a damn fool, and he was going to get a Nobel Prize," House says, still visibly upset by it fifty years later. "It was my first disappointment of many that followed."

• • •

News of what House had done reached an auditory researcher named Blair Simmons, who was a new assistant professor of otolaryngology at Stanford University School of Medicine. Simmons was interested in the physiology of sound reception. His studies of the auditory systems of cats had been published in the prestigious journal *Science*. A scientist's scientist, Simmons was angry at what he regarded as "irresponsible claims" emanating from Los Angeles, but he was captivated by what he considered a true research problem.

A year later, in 1962, Simmons saw an opportunity. An eighteen-year-old cancer patient with increasing hearing loss was going to have exploratory brain surgery under local anesthesia. As with House's patients, the young man's inner ear would be exposed during the operation. In his work with cats, Simmons had successfully implanted electrodes into the inner ear without destroying the auditory nerve. Now he wanted to try his technique in a human subject who would be able to describe what he heard. "We were amazingly lucky on our first try," Simmons later wrote. With an electrode stimulating his auditory nerve, the teenager heard a wide variety of sounds. Most surprising was the boy's ability to hear sounds at either end of the spectrum beyond the range required for speech.

Two years later, Simmons went a little further. Sixty-year-old Anthony Vierra of San Jose suffered from retinitis pigmentosa, a condition that causes an increasing loss of peripheral vision. Eventually, the tunnel vision narrows completely and the patient is left blind. By the time he met Simmons, Vierra was also profoundly deaf in his right ear and was losing what hearing remained in his left. He gamely agreed to be permanently implanted with a six-electrode cochlear implant that Simmons had devised. Like the House implant, this one had wires threading through the skull, just behind the ear, that had to be connected to a computer or electronic stimulator in the laboratory for Vierra to hear anything. The surgery was performed at Stanford, but even a respected researcher like Simmons had trouble finding basic-science colleagues willing to work with him on testing such a controversial project. So he turned to one of the few places in the country with an established interest in the alchemy of electricity and human speech: Four scientists at Bell Labs, which had moved from Manhattan to New Jersey, agreed to perform the audiological testing of Vierra. "They were outsiders," Simmons commented later. "I don't think they'd read the publicity." Vierra had never been on a plane before, and since he was nearly blind, Simmons and his wife escorted him across the country.

The combination of Vierra's poor vision and hearing meant the researchers had to communicate with big block-letter signs saying things like: TELL US WHEN YOU THINK YOU HEAR SOME-THING. Vierra needed some lessons in how to accurately describe and compare the sounds he was hearing, but he ultimately was able to identify them in terms anyone could understand. At a slow rate of one pulse per second, he heard a ping or a ding. Three to four pulses per second resulted in clicks. As the rate increased, he heard a buzz, then a sound like a telephone ringing, and finally a car horn above thirty pulses per second. Vierra was able to recognize familiar tunes like "Jingle Bells" and "Mary Had a Little Lamb." In the eighteen months he wore the device, however, he never could understand speech. In a *Science* article on the work coauthored by Simmons, his Stanford colleague John Epley (who collaborated on the surgery), and the Bell Labs researchers, the conclusions were cautious. They had succeeded in expanding technical knowledge about pitch perception, although in a qualified way. As to the larger goal, they wrote: "Much remains uncertain. . . . It is unlikely that stimulation with any speech-derived signal would permit this subject to discriminate an appreciable number of words, unless considerable learning were possible."

Early in 1967, Simmons presented everything that was known about cochlear implants at a Chicago conference on microsurgery of the ear. "I am glad this meeting is a workshop," he began, "because most of what I have to suggest means exactly that: work." The auditory stimulation done to that point had been crude, said Simmons. "We must be able to produce an orderly and predictable array of pitches, loudnesses, rhythms, etc. These must be close enough to normal neural patterns so that a deaf person's task in learning will be no more difficult than learning a foreign language." Should they ever succeed, Simmons guessed it would take new users at least as long to acclimate as it takes babies to decipher the auditory world around them: a year, give or take a few months.

Success would require a cooperative effort, Simmons said. Part of the problem was just how little was known about how the brain understood pitch. Frequency information was transmitted in two ways: via the rhythm of the sound, technically the repetition rate of the stimulus; and via the place on the basilar membrane that is tuned to respond best to a particular frequency. Like real estate investors, scientists were essentially asking which mattered more: timing or location. Simmons was blunt about their ignorance. When he displayed Mr. Vierra's responses to variations in the rate of the pulse, the lines of the graph lay on top of a photograph of a crater of the moon. "The exact information we have about the major portion of the conventional speech frequencies (500–2,000 Hz) is best represented by the rather large hole I have placed in the background," he said. The question of pitch needed an answer because it would guide the placement of electrodes and decisions about what information to send over them. Even if no one ever succeeded in building a workable artificial ear, Simmons concluded, the knowledge they would gain about how the ear and the brain process sound would be valuable and important. Besides, he finished a little impishly, a cochlear implant "just *might* be possible."

Bill House had "unabashed admiration" for Simmons. It encouraged him that someone of the Stanford researcher's stature was tackling cochlear implants. "You have no idea how many people have told me that this problem is completely unsolvable," House told Simmons that day in Chicago. But he disagreed about the need for animal studies and working out answers in the laboratory—standard operating procedure for most research scientists. "We must be willing to take some risks in applying what we have already found out," said House. He insisted there was benefit to patients even "if they can only hear sounds such as footsteps and auto horns."

Anything House could offer appealed to Charles "Chuck" Graser. The California man had been writing to House every six months since news of the 1961 surgeries broke. A high school teacher, Graser drove a tanker truck in the summers to make extra money and had been badly burned in a truck fire. Doctors had given him streptomycin to fight infection, and it had caused him to lose his hearing. In the ten years that he was deaf before getting an implant, Graser depended completely on lipreading and he experienced "strictly silence." He had begun to think, he joked, that in order to communicate he would have to carry around a pair of scissors to cut off all the newly fashionable mustaches. After the accident, he had to give up his teaching job and was only able to take on part-time work as a school librarian. Like many people who had lost their hearing as adults, Graser experienced deafness purely as a loss of the life he had once had.

Nearly ten years after they began corresponding, Bill House finally told Chuck Graser he was ready to try again. Biocompatible silicone had been developed by the inventors of the pacemaker, and in aerospace engineer Jack Urban he had found a collaborator who was more of a soul mate than Jim Doyle had ever been. Most of Urban's other work was defense-related, and he told House he was "no longer anxious to help blow people up" and wanted to do some good. That suited

House, who never applied for a patent for his implant because "I felt it would restrict others who might want to pursue this promising lead." Since both men had day jobs, they hammered out ideas over dinner in their regular booth at a favorite Italian restaurant.

Like the earlier prototype, the prosthesis that House and Urban built for Graser and two others could only be used in the laboratory. An early computer served as both microphone and sound processor and sent its instructions along wires that connected to the implanted electrodes via the metal plug behind the patients' ears. Of the volunteers, Graser proved to be the most determined and interested. He came to Urban's laboratory several days a week and kept extensive field notes of his own. "We knew exactly what electricity was going into the head because he was hardwired," says House. "It allowed us to try all different kinds of electrical stimuli. Then we'd think of something else. Everything we built was battery-operated. We weren't going to hook him up to the wall circuit because we didn't know what was going to happen." As an additional precaution against the delivery of too great a shock, they had Graser sit on a wooden chair placed on a rubber mat. Two years into the work, in 1972, Urban miniaturized the electronics and packaged them in a gunmetal-gray box the size of two stacked decks of cards, which Graser could wear on his belt or in a pocket. House worried about the untested effects of constant stimulation of the nerve, but they were all excited by the prospect of allowing Graser to hear at home.

Suddenly, Graser could hear a dog bark and was able to recognize the squawk of blue jays from time to time. He wrote of his experience: "You would probably describe my current progress as changing from profoundly deaf to just hard of hearing, but difficulty hearing and comprehending is in a completely different league from silence. For instance, tonight I can finally hear the bell that indicates that I am at the right hand margin, as I type this letter. . . . I used to be a [ham] radio operator, and sometimes I would get a distant signal that I couldn't really hear. It sounded dim and garbled. That's the way this sounds. It's definitely an electronic sound."

Graser and his wife, Barbara, took to playing games to practice his listening skills. House's daughter, Karen, a filmmaker, captured one for posterity.

"Say some of the simple words to get me started," says Graser, a square-jawed, black-haired man, who was forty-four at the time. For the game, he removes the electronics case, with its microphone and processor, from his pocket and holds it up close to his wife's mouth. Then he looks away so he can't see her lips.

Barbara leans in close with her arm resting on his shoulder and says, "Baby."

"Say it again."

"Baby."

"Baby," he repeats. Then he turns to look at her. "Am I right?"

She nods and he turns away again. Concentrating hard, Graser gets "sailboat" and "earthquake" but stumbles on "cowboy." Then they try some sentences. He misses "It's time for supper." But the next one is easy: "Baby, it's cold outside," he repeats after her with a big grin. "I remember that one."

Listening to the world now, especially to music, is a bit like the game he plays with Barbara, he says: "You put enough clues together, and you hope you're pretty close to knowing what's going on."

On the same film, a nineteen-year-old woman named Karen, the first congenitally deaf person House implanted, is shown hearing sound for the first time. She is sitting in the laboratory with machines all around her. Unlike Graser's, her electrodes had been completely implanted. The behind-the-ear plug was gone, and signals were to be sent via radio transmission through the skin from a coil held to the outside of her head. Until the external and internal coils lined up, there would be no sound. (In later versions, a magnet solved this problem.) In the clip, House stands behind Karen, holding back her long dark hair and circling the external coil slowly behind her ear. "Can you hear my voice?" Jack Urban says periodically. Karen's sister, Andy, stands nearby, echoing Urban: "Can you hear my voice?"

Suddenly, Karen's eyes widen and her face brightens. She raises her closed fist and shakes it forward and back, signing: YES! YES! Her sister signs that she should begin counting . . .

"One . . . two . . . three . . ." Karen's voice has the flat, nasal tone common to the deaf. She shakes her fist again, then hesitates as the sound disappears.

"One . . . two . . . three . . . four . . ." Her eyes widen again, and she begins to cry. She holds her hand to her chest for a second, then covers her face momentarily. It is the first time she has ever heard her own voice.

A few minutes later, Karen is showing off. The external transmitter has been fixed in place with a headband. Urban is playing "Six Variations on the Turkish March" by Beethoven, and she is grandly rolling her hand away from her mouth to indicate someone singing, to show that she recognizes music.

"It was a moment I will never forget," House tells me as we sit watching the film together.

"Did you get emotional yourself sometimes?" I ask.

"Yes," he says quietly.

Parts of the movie of Karen were shown on *60 Minutes*, *Nova*, *That's Incredible!*, and a 1975 National Geographic special called "The Incredible Machine." The dreams of science fiction seemed to have been made real.

As Chuck Graser proved, some people did very well with the House implant. But the device had serious limitations. Based on Graser's responses, House and Urban had decided to use only one electrode— a single channel of information—that would stimulate the entire cochlea at once rather than separate electrodes capable of stimulating discrete sections of nerve fibers. While there was a lot researchers didn't know about hearing, House's device violated one principle they did know: the tonotopic theory, the idea established by Békésy that frequencies were laid out along the basilar membrane like keys on a piano. The Frenchmen—Djourno and Eyriès—had already concluded that multiple channels would be necessary to understand speech. Dr. Nelson Kiang, one of the most prominent auditory neurophysiologists in the nation, maintained that, Graser's experience notwithstanding, a single-channel implant could only produce a kind of buzzing. "Enthusiastic testimonials from patients cannot take the place of objective measures of performance capabilities," said Kiang. Scientifically speaking, he was right. "There were some basic fallacies in House's understanding of auditory physiology," Marc Eisen told me. Yet it was undeniable that something was working for patients like Graser.

"I remember once having an argument with [House] about the sort of simpleminded engineering approach he had adopted," Michael Merzenich told me. "And he said, 'Mike, if I have a man that has no leg and I can give him a stick to prop his leg up on, I'm going to give it to him.' The first thing I thought was, That's ridiculous. Then when I thought about it a little bit, I thought, Well, you know, so would I." Merzenich laughed heartily at the memory. "You have to give him a lot of credit for being courageous. He was a man filled to the brim with gumption."

In 1975, all thirteen existing cochlear implant patients—each with a single-channel device—traveled to Pittsburgh for a week of evaluation. The National Institutes of Health had waded into the debate and funded an independent review headed by Dr. Robert Bilger, an audiologist and neurophysiologist at the University of Pittsburgh. Eleven of the patients, including Graser, were House's and two had been implanted by a San Francisco doctor named Robin Michelson, who had gotten into the game a little later than House and Simmons and created a device of his own. (It was Michelson who brought Merzenich into the work.) Bilger began the study as a skeptic but emerged modestly optimistic. His report was released in 1977, and the findings were announced by the chairman of Pittsburgh's ear, nose, and throat department, Eugene Myers. There was no longer any question that cochlear implants worked, said Myers. They did. Users found them helpful with lipreading and had improved awareness of environmental sounds. Myers likened their stage of development to the Wright brothers' flight at Kitty Hawk. Bill House had indeed reinvented the airplane.

It should have been vindication, yet the moment was bittersweet. Yes, cochlear implants worked, said the report, but House's single-channel device was not the way forward for understanding speech. Overall, the average closed-set word recognition score in the Bilger study was 16 percent (closed-set word lists are related and somewhat predictable)—better than without an implant but a long way from conversational understanding. The study concluded: "[A single-channel device] will not provide a speech input that either sounds speech-like, or is understandable. . . . Until such multichannel prostheses become a reality, one must consider the question of whether or not it is reasonable to continue implanting single-channel prostheses."

House was frustrated and angry. Perhaps this is the saddest part of his story, the part that explains how House could be viewed as both a hero and a kook. For decades after the Bilger report, House defended his single-channel device, insisting that it did allow some people to hear speech. That was true, but only for a very few. "The single channel provides an electrical field without any pitch discrimination," explains Dr. Simon Parisier, one of the first otolaryngologists to embrace cochlear implants. "Most people get good information [combining it] with lipreading; a few star people could understand without lipreading." But for the majority, a single-channel implant provided

9

PRIDE

In 1967, the same year that Blair Simmons stood in front of a crowd of doctors and scientists to discuss the possibility of creating an artificial ear that worked, another man took to a different stage one thousand miles to the east for a presentation that would help to transform the world of the deaf in a completely different way.

"The lone actor strutted proudly to the center of the stage, chest out, and head held high," wrote Jack Gannon in his history of the deaf in America. "He bowed slightly, then straightened and moved his head slowly sideways as his eyes scanned the audience. He raised his right, white-gloved hand and fingerspelled with machine-gun rapidity: 'G-i-a-n-n-i-S-c-h-i-c-c-h-i.' He clapped his hands and suddenly the stage was overflowing with colorfully-costumed actors and movement . . . movement everywhere. Hands, fingers, faces, bodies, and voices began to communicate. No one had ever seen anything like it before."

This unusual version of Puccini's comic one-act opera was one of four short plays that made up the first performance of the National Theatre of the Deaf, and though only six people bought tickets, the curtain had been raised on a new era.

The idea for a theater company showcasing deaf actors was first proposed in the 1950s by Edna Levine, a psychologist who worked with the deaf, and Anne Bancroft and Arthur Penn, the star and director respectively of the Broadway production of *The Miracle Worker*. They couldn't get funding and the project died. In the 1960s, the idea was revived by Broadway set designer David Hays, who had been captivated by a signed performance he saw at Gallaudet. Hays joined forces with Bernard Bragg, a deaf actor whose popular one-man mime performances earned him a four-year run on television in a show called *The Quiet Man*. They found a fitting home for the experimental troupe in the newly established Eugene O'Neill Theater Center in Waterford, Connecticut.

After oralism took over in deaf schools at the end of the nineteenth century, sign language was relegated to "a villain-like role," wrote Gannon. Teachers and parents blamed sign language when deaf students struggled with English. "As a result, sign language became unpopular and a stigma was attached to it which made many [people] uncomfortable and unwilling to use it in public."

Ted Supalla, a linguist who studies ASL at Georgetown University Medical Center, and who also happens to be married to neuroscientist Elissa Newport, grew up in such an environment. In the Deaf world, Supalla is "deaf of deaf," meaning his parents were deaf and he was raised signing. Even if he wasn't an expert on ASL, that makes him royalty of a sort, at the top of the Deaf social hierarchy. It was his younger brother, Sam, who had to be told the little girl next door was HEARING.

Although the Supallas used sign language at home and at the Deaf clubs where they socialized, Ted, who is profoundly deaf and got no help from hearing aids, began school in the 1950s and had to communicate with the teachers at the Oregon School for the Deaf in English. When we met at Georgetown, I could see that his ASL was beautiful, though I had to rely on an interpreter because mine was rudimentary. He described a day at school when he was clearly frustrated and a teacher took him out to the hall to talk to him alone. To his amazement, she began to use sign language. His relief was so enormous, he remembers, "I started to cry." But then he had to go back into the classroom and use his voice, which was not something he did well. "I just read books in class all day long," he says. "I knew I was missing out." His solution was eventually to transfer to a hearing school, which he attended without the benefit of an interpreter as there were none to be had in those days. If he was going to be hamstrung by lack of communication either way, he figured he was better off in the hearing school.

Then one day, as a young man, he attended a performance of the fledgling National Theatre of the Deaf, who "put sign language onstage literally," he says. For Supalla, that moment led to a career, first in the theater and then studying sign language as an academic. All around him, other deaf people were having a similar experience and found themselves thinking about deafness and sign language in new ways.

That had been the goal. When Hays and Bragg and their colleagues started work, they took an entirely new approach. Although pantomime was popular in deaf theater groups in Europe, and it was what Bragg was known for, the group chose to limit its use in their productions. "They decided to make NTD a theatre of language—a theatre that would concentrate on visual language," wrote Gannon. For *Gianni Schicchi* and all subsequent performances, they brought theater to a deaf audience who could, of course, follow the signing (it was initially a theatrical version of signed English). But they always included hearing actors who signed and spoke their own roles and interpreted for the deaf actors, thereby making signing accessible to the hearing. Not every critic loved every performance, but some plays were immensely successful. *Boston Herald* drama editor Samuel Hirsch described the result as "pure art, drawn from a new medium of human expression." The critic for *The National Observer* called the shows "exciting, inventive, beautiful, and unusual."

In 1971, NTD moved beyond translating existing works when cast members produced their first original play, *My Third Eye*. It included a biographical segment in which the actors described some of their experiences as deaf children. Bragg told of being taken away to a residential school at the age of four and a half. As he and his mother, who was also deaf, traveled to the New York School for the Deaf, she didn't reply to his urgent questions about where they were going. Once there, they stood in the superintendent's office.

"I was filled with the sickening panic of a washed-away world. My mother kissed me and said, 'This is the place where you will get all your education . . . ,' she kissed me and was gone."

Mary Beth Miller said: "I would teach some of the kids who knew no sign language. The teacher would tie my hands together."

Ed Waterstreet: "Same in my school, when I tried to sign with my friends the teacher caught me and made me sit on my hands."

That they could now be using that same language onstage was an extraordinary turn of events. As Bragg put it: "People used to push my hands down in embarrassment and tell me not to sign in public. Now people pay to see me perform in sign language."

The change in perspective did not just spring from the minds of these actors. Others had begun to think that sign language deserved more respect. Chief among them was a Gallaudet English professor named William Stokoe. He was hearing, as were many Gallaudet professors in that era, and was unfamiliar with sign language until he got to campus. Fascinated, he enlisted the help of two deaf assistants, Carl Croneberg and Dorothy Casterline, and began filming people as they signed. As he pored over the resulting films, Stokoe began to see patterns.

Until then, the assumption was that sign language wasn't a language at all. Beyond being a means of communication, languages had to have, in linguistic terms, rules governing the way their symbols could be used, combined, and transformed. No one had ever looked at sign language this way or seriously considered that it might qualify. In thousands of hours spent studying people as they signed, Stokoe found rules: units of meaning, patterns of word order, and points of contrast. "He was the first linguist to subject sign language to the tests of a real language, and he found that it withstood them all," wrote Gannon.

At first, no one was impressed. Even deaf people were skeptical or dismissive. According to Carol Padden and Tom Humphries, the thinking was that communicating in signs was "just something we did." Many had absorbed the negative view of what they called "the sign language." In 1965, however, Stokoe, Croneberg, and Casterline got more attention when they published *A Dictionary of American Sign Language on Linguistic Principles*. The book marked the first time the sign language was given the formal name American Sign Language. (Sign languages are no more universal than spoken languages, and each country has its own. Even English-speaking countries such as Britain and the United States use very different sign languages.)

Instead of describing signs according to pictures, the authors recognized that the signs were composed of smaller parts—such as handshape, location, and movement—akin to the phonemes and morphemes (units of sound and meaning) in spoken language. In the *Dictionary*, signs were organized according to those smaller parts rather than categories or English translations. Signs in which all five fingers of the hand were spread in the 5 sign, for example, were listed together according to their location: on the chin (MOTHER), on the forehead (FATHER), and on the chest (FINE). Add movement and you get GRANDMOTHER (by bouncing the hand away from the chin) and GRAND-FATHER (bouncing it away from the forehead). Stokoe even invented a complicated new system of notation to describe each sign. Though the response to Stokoe's work was mixed at the beginning, he had planted the seeds for what quickly grew into a science of signed languages.

Meanwhile, the civil rights and feminist movements were unfolding through the 1960s and 1970s, and new ideas of identity and political consciousness began to take hold. By the 1984 publication of Harlan Lane's history of the deaf, *When the Mind Hears*, that thinking had matured. The book, from which I learned so much about early deaf education, was a call to arms. Lane called it "a study in the anatomy of prejudice" and claimed that a fear of diversity had led the hearing majority to oppress the deaf minority over the centuries through a series of "sustained outrages against fundamental human values" such as "the attempt to force assimilation, to claim biological insufficiency when assimilation fails, to indoctrinate minority children in majority values through the schools." The deaf, declared Lane, do not belong to a class of disability but to a linguistic minority.

Most of the deaf were far less politicized than Lane, but he was right that, like many minorities, they had long found comfort in each other. They knew they had a "way of doing things" and that there was what they called a "deaf world." Largely invisible to hearing people, it was a place where many average deaf people lived contented, fulfilling lives. Lou Ann Walker depicts it beautifully in her memoir of growing up with deaf parents, *A Loss for Words*. In her father's work as a newspaper printer, with their deaf friends, and at home with their three hearing daughters, Gale and Doris Jean Walker were loving, intelligent, and capable people. When their deaf world collided with the hearing world, however, they nearly always came away bruised. Walker describes hearing relatives who didn't want to be seen in public with her parents. Her non-signing maternal grandfather attempted late in life to tell his daughter he loved her, but afterward Doris Jean asked Lou Ann what it was her father had said. Children of deaf adults

(CODAs) like Lou Ann Walker are uniquely able to pass between the two worlds, but they absorb some of the pain of their parents with each border crossing. Near the end of her book, Walker describes interpreting the funeral of that same grandfather for her parents:

> As I signed the service, I tried not to think of the words on my hands. I didn't want to get emotional as I conveyed my grandfather's funeral sermon to my mother and father.
>
> Standing there, taking the invisible words from the air and placing them for my mother and father to see, I searched my mother's face for echoes of the face of the man lying in the box a few feet behind me.
>
> "Ashes to ashes, dust to dust. We come here to bury Chester Cooper Wells." I signed "bury" with two open hands lowering him straight down into the earth—but gently, gingerly. I didn't want to hurt my mother any more than I had to.

No one had ever tried to name the world that people like Gale and Doris Jean Walker lived in. They were part of a grassroots deaf community. Beginning in the 1980s, however, deaf people, particularly in academia and the arts, "became more self-conscious, more deliberate, and more animated, in order to take their place on a larger, more public stage," wrote Padden and Humphries. They called that world Deaf culture in their influential 1988 book *Deaf in America: Voices from a Culture*. The capital "D" distinguished those who were culturally deaf from those who were audiologically deaf. "A large population, established patterns of cultural transmission, and a common language: these are all basic ingredients for a rich and inventive culture," they wrote.

"The traditional way of writing about Deaf people is to focus on the fact of their condition—that they do not hear—and to interpret all other aspects of their lives as consequences of this fact," Padden and

Humphries wrote in the introduction. "Our goal . . . is to write about Deaf people in a new and different way. . . . Thinking about the linguistic richness uncovered in [work on sign language] has made us realize that the language has developed through the generations as part of an equally rich cultural heritage. It is this heritage—the culture of Deaf people—that we want to begin to portray."

It was a time of soul-searching within the deaf community. The hearing population got a taste of the internal debate when Marlee Matlin won an Academy Award in 1987 for her performance as a troubled young deaf woman in *Children of a Lesser God*. For the deaf to have one of their own so honored and to have her sign her acceptance speech to millions of viewers was exhilarating. But it came as a shock to Matlin and some of the hearing observers when she was criticized for using her voice (something she rarely does) for part of her acceptance speech.

The message of the emerging deaf civil rights movement was this: Deafness is not a disability; it is a difference. It is not something to be "cured" or "fixed." It is even something of which a person can be proud. The new consciousness found its greatest, most powerful expression in the Deaf President Now protest at Gallaudet University in March 1988.

WE STILL HAVE A DREAM! read the banner. It was so big it took up most of a city street and required more than thirty people to carry it. In a nice bit of historical solidarity, the banner had been borrowed from the Crispus Attucks Museum in Washington, DC, and had last been used to rally for a national holiday in honor of Martin Luther King Jr. But this time, on March 11, 1988, it was carried by deaf students from Gallaudet University. Although the rounded dome of the Capitol Building is visible from the upper floors of buildings on the Gallaudet campus, the students were borrowing another tradition from previous civil rights protests and marching on the Capitol as the culmination of a week of tumult and emotion. They were now carrying their concerns beyond their own community and out to the wider world.

All week, the students had demanded that Gallaudet's board of trustees appoint a deaf president. In its 124 years of existence, that was something the university had never had. Over time, more and more deaf administrators and faculty had been appointed, but the president and most of the board members, including the chair, Jane Bassett Spilman, were hearing. What's more, Spilman and most of the hearing trustees had never learned to sign. In the eyes of the Deaf community, that lumped them squarely in the paternalistic tradition of hearing people who think they know what's right for deaf people without having any idea what being deaf is all about. The notion that deaf people were not capable of managing their own university was deeply offensive.

When the previous president, Jerry Lee, stepped down in 1987 and a search committee was formed, members of the Deaf community had made clear their preference for a deaf president. By early 1988, they were sure it was their moment. "It's time!" read an invitation to show solidarity at a rally on March 1. "In 1842, a Roman Catholic became president of the University of Notre Dame. In 1875, a woman became president of Wellesley College. In 1886, a Jew became president of Yeshiva University. In 1926, a black person became president of Howard University. AND in 1988, the Gallaudet University presidency belongs to a DEAF person."

But on Sunday night, March 6, the trustees announced in a press release that Elisabeth Ann Zinser, the only remaining hearing candidate, vice-chancellor of the University of North Carolina at Greensboro, would be the next president. Shock, anger, and disbelief raged through campus. An angry crowd of students, faculty, and alumni set fire to copies of the press release and then, gathering outside Gallaudet's main gate on Florida Avenue, they decided to march into the center of Washington to the Mayflower Renaissance hotel, where the board meeting had been held, to demand an explanation from Spilman.

When she finally met with the student leaders that night, they understood her to say, "Deaf people are not ready to function in the hearing world." She later maintained that she had not said that and never would have, and that her interpreter had misunderstood her. But the damage was done. The statement fueled a fire of frustration and resentment that had been smoldering for years. A few days later, Spilman again embodied the problem when she stood at the podium at Gallaudet's field house as angry, jeering students refused to pay attention.

"We will not sit here if you are going to scream so loudly that we cannot hear you and that we cannot establish a dialogue," said Spilman.

As her words were interpreted, the students replied that there was no noise for them. Later they set off the building's fire alarms, just for good measure.

By the end of the week, the demonstrations had succeeded completely. Elisabeth Zinser resigned before she ever set foot on the campus, and the board appointed I. King Jordan, a deaf psychology professor, as the new president.

The protest had been heard far beyond the gates of the university and the meeting rooms of the Mayflower hotel. On March 11, the day of the march on the Capitol, ABC News named the Gallaudet student government president, Greg Hlibok, as its Person of the Week.

"This week, millions of us have had the chance to sit forward in our chairs and watch a group of young people as they break down some very out-of-date stereotypes," said ABC's anchor Peter Jennings in his introduction.

Then Hlibok, a slim, clean-cut young man in a jacket and tie, appeared. "Deaf people are capable of anything except that they can't hear," he said through an interpreter.

Jennings recounted the events of the week, the "test of wills" that had brought down a university president and raised the national consciousness for hearing people. For deaf people, he said, it amounted to "a national catharsis."

"It seemed as if there'd never been such an opportunity," said Jennings, "to tell the world that deaf does not mean defeated." For the first time ever, it was hearing people who had come away from a collision dented.

Into the turbulence of nascent deaf civil rights dropped the cochlear implant.

10

LANGUAGE IN THE BRAIN

It was about eight months after Alex got his hearing aids and started at Clarke, and Mark and I were sitting anxiously in an exam room at New York Eye and Ear Infirmary while Alex, now two and a half, played with the pale green vinyl footrest of the exam chair. We were waiting for Dr. Simon Parisier. Medically speaking, we had graduated to the big leagues.

"We need to implant him," Dr. Parisier said soon after he came in.

A few months earlier, at the beginning of September, Alex and I had returned to Jessica O'Gara for what we thought was a routine appointment to fine-tune the settings on his hearing aids. We got an unwelcome surprise: All of the hearing in his right ear was gone. He was now profoundly deaf on that side. Through the summer, we had watched obsessively for bumps on the head that might worsen his hearing, but nothing had happened. Or so it had seemed. Perhaps, as Dr. Dolitsky had warned, Alex's remaining hearing had been lost from something as seemingly minor as a sneeze. If the right ear had dropped so precipitously, it was likely only a matter of time before the left ear followed. Then came a speech and language evaluation at the beginning of December—a year after the first one. Alex had moved from the second to the sixth percentile for receptive language and was still in the eighth percentile for expressive language.

"Single digits? Still?" I gasped when I got the results. "But he has more than two hundred words now! That can't be right."

It was right. Alex had been working hard and learning every day. But so was every other two-year-old. No one was going to stand still and wait for him. That is why achievement gaps are so stubbornly hard to close. The learning curve for children who start behind can't simply parallel that of typical children. To catch up, their progress requires the steep trajectory of a rocket. Alex's new drop in hearing was at least a reasonable explanation for why he was moving slowly. His right hearing aid was no longer providing useful sound. He was down to one ear, and a limited one at that.

Immediately, Dr. Dolitsky had sent us to see Dr. Parisier, one of the pioneers of cochlear implant surgery. As a younger doctor in the 1970s at Mount Sinai Hospital in New York City, Parisier had embraced the potential of the new technology from the start. Back when he graduated from medical school in 1961, the introduction of antibiotics had just radically changed the field of ear surgery. Before that, he says, "it wasn't uncommon to die from an ear infection. In the 1950s, doctors hadn't been concerned about hearing, but with saving lives. There was nothing you could do about nerve deafness." With the introduction of implants, Parisier felt there might be something he could do to help previously unhelpable patients. He found meningitis survivors particularly compelling. "Meningitis was the single largest cause of deafness in kids," he told me. "Ten percent of those who survive become profoundly deaf. It was very painful. A child goes into the hospital able to speak at two or three years old and within days, that speech goes. When they wake up, they don't talk anymore." Even six- and seven-year-olds struggled and lost their language. The experience taught Parisier about the importance of getting language early. It worried him that Alex was already two and a half.

A new round of tests ensued, this time to see if Alex was a candidate for a cochlear implant. Mark and I began to investigate what this would mean. At Clarke, which serves children through age five at the New York campus, I already knew many kids with implants. Some spoke intelligibly, some much less so. I knew the device didn't bring instantaneous success, but also that Alex would be starting with the advantage of having some hearing in the other ear.

We brought home videotapes prepared by the three cochlear implant manufacturers. Of course, they showcased the success stories. But what stories they were. In the "before" videos, the children were Alex's age, laboring with speech therapists to pronounce a few words. In the "after" videos, they were school-age, in mainstream classrooms, all of them speaking, some of them virtually indistinguishable from their hearing classmates. One particularly captivating profoundly deaf girl had used sign language until she was three, when she got her implant. By the movie's end, she was nine, a busy, happy fourth-grader, singing in her mainstream school's choir. As the credits on the video ran, Mark and I looked at each other.

"That's what I want," he said.

"Me too," I agreed.

We emerged with a case of emotional whiplash: We went from desperately wanting to hang on to every decibel of Alex's hearing to desperately hoping his hearing was bad enough to qualify him as a candidate.

White-haired and soft-spoken, Dr. Parisier had a gentle, matter-of-fact manner, but he didn't mince words when he delivered his opinion. He clipped several CT scan images of Alex's head up on the light board and tapped a file containing the reports of Alex's latest hearing tests and speech/language evaluations. "He is not getting what he needs from the hearing aids. There's no high-frequency access. His language is not developing the way we'd like," he said. Then he turned and looked directly at us. "We should implant him before he turns three."

A deadline? So there was now a countdown clock to spoken language ticking away in Alex's head? What would happen when it reached zero? Alex's third birthday was only a few months away.

As Parisier explained that the age of three marked a critical juncture in the development of language, I began to truly understand that we were not just talking about Alex's ears. We were talking about his brain.

Even more than his hearing, Alex's brain was mysterious. It required faith. I couldn't touch it or caress it, as I did his ears, running my fingertips along their small curving lobes. Of course, I knew his brain was there. When he breathed or blinked or blew me a kiss, it was sending proof of its existence, like postcards from a far-off land I can point to on the globe but never visit. Still, I struggled to imagine the reality of it. Even the vocabulary of the brain is hard to fathom with its map of place names that sound simultaneously alien and ancient— superior temporal gyrus and perisylvian cortex, amygdala and hippocampus.

In among the models and diagrams of the ear in every doctor's office we had visited, there was not one map of the brain on the wall. Perhaps there should have been. I wanted to know more than that the cochlear implant would deliver sound to Alex's auditory nerve, which would convey it to the brain. I needed to know what would happen to that sound once it arrived. I needed to know what the brain would do with that sound while Alex was two that it might not be able to do once he turned three.

The answers lay in the burgeoning science of brain plasticity—the study of the capacity of the brain to change with experience. Helped by advances in technology, neuroscientists have built up atlases of normal brain development. That is the starting point for understanding what happens in brains that are atypical.

The differences between a brain at birth and that same brain at twenty-one years of age are considerable and driven both by genetics, which is "merely an opening gambit on nature's part," as science writer Sharon Begley put it, and by environment. The baby's brain arrives in the world primed for learning. It is now thought to contain seventy to eighty billion neurons, or nerve cells, down from previous estimates of one hundred billion, but still most of the neurons a person will need as an adult. What is lacking is communication from neuron to neuron. The process of generating the remaining necessary neurons and creating connections between neurons is what the learning of childhood is all about.

Neurons are often compared to trees, with roots and branches, but to me they look more like spindly sea creatures or insects with long skinny bodies and arms like tendrils waving away at each end. Information comes in through those tendrils, the dendrites, and travels along the axon to the other end of the cell. At the base of each axon there's a gap before the next neuron. The electrical signal is carried over that gap, or synapse, by chemical neurotransmitters, like a written message wrapped around the shaft of an arrow and fired over a moat.

The warmth of a baby blanket, the smell of a father's aftershave, the sight of a mother's loving gaze, the soothing melody of a lullaby: Together, these perceptions contribute to a baby's sense of security, but each is also altering the configuration of the baby's brain. Each is a piece of sensory information that travels through the nervous system to the appropriate area of the brain. The image of a mother's face, for example, enters the retina of the eyes and passes along a pathway to the visual cortex in the occipital lobe at the back of the brain. Every time the baby sees his mother's face, the same chain of nerve cells fires. As neuroscientists say, neurons that fire together wire together. Pretty soon, a circuit has been created and the baby recognizes his mother instantly. There are similarly specialized areas for touch, sound, and smell.

In this manner, information travels through the brain leaping from neuron to neuron, creating circuits that receive and then process and act on information. Zip, zip, zip. The more often a specific signal is sent down a particular path, the more defined and efficient that path becomes, like a hiking trail that is well maintained versus one that's overgrown. Paths that are traveled less often or not at all—connections that aren't made, in other words—eventually disappear. The brain assumes they are unnecessary and prunes them away. Snip, snip, snip. As neuroscientist Helen Neville puts it, "experience is like a sculptor who begins with more clay than he needs." This idea is somewhat counterintuitive—that a child's brain is simultaneously strengthening and eliminating brain circuits. But that is the beauty of the system. It is as if the landscape of the brain is under the control of a master gardener who creates paths where they are wanted, plants new seeds as necessary, directs the flowerings and branchings, and cuts back to concentrate growth where it is most desirable.

Today, scientists can watch the brain maturation process unfold. In a prospective study published in 2004, a group of researchers based at the National Institutes of Health and UCLA followed thirteen healthy children for close to ten years. The children, who ranged in age throughout the study from four to twenty-one, underwent functional magnetic resonance imaging (fMRI) every two years. The images were color-coded to indicate relative levels of gray matter density in the children's brains, with red and yellow indicating more gray matter and blue and purple less. As neural connections explode early in childhood, gray matter density increases. As those connections are pruned back and insulated with myelin, that gray matter volume decreases and white matter increases. So more white matter indicates a more mature brain.

In the time-lapse movie of the results, large swathes of yellow and red are slowly washed with blue as the exuberant bursts of neuron growth in early childhood become cooler and more efficient, mature neural pathways. The sequences showed that the brain's most basic functions, those concerning vision and touch, developed first, followed by centers for language in late childhood and early adolescence and then, toward early adulthood, by the areas that control reasoning, self-reflection, and planning. (Interestingly, in Alzheimer's patients the sequence is reversed. Their brain matter loss progresses from front to back, beginning with the higher-order functioning and finally enveloping more basic functions of the brain.)

Most of the growth in neurons takes place between birth and the age of five, which is why the UCLA movies, whose youngest subjects were four, begin with brains full of gray matter (i.e., uninsulated neurons). The pruning begins in earnest around five, and children will lose 35 percent of the nerve cells they have at five before adulthood. For many years, the assumption was that after early childhood our brains did not continue to generate new nerve cells, but within the past ten years, researchers discovered a second, smaller wave of neuron creation in puberty. And we now know that it is possible to keep learning—and even generating new neurons—into adulthood, though it takes far more effort to see results.

According to the brain's basic division of labor, sensory areas receive all the incoming information from our eyes, ears, nostrils, tongue, and fingertips. The back of the brain, the occipital lobe, handles vision. Hearing is processed in the temporal lobe on the side of the head just above the ear. A strip across the top of the head collects information about everything we touch, whether with our toes, palms, shoulder, cheek, or any other part of the body. Motor areas control our movements. The limbic system is involved with emotion. "Cortex" is the collective name for the exterior surface of the brain, the wavy, lumpy gray matter familiar from anatomy books. It's here, particularly in the section right behind the forehead known as the prefrontal cortex, that the higher-order processing occurs that marks us as distinctly human and controls our ability to plan, to reason, to remember, to reflect.

The central auditory system begins where the inner ear passes a signal to the auditory nerve. A part of the eighth cranial nerve, the auditory nerve is not just one nerve but a bundle of nerve fibers—a coaxial cable of sorts—connecting the cochlea to the brain stem. Within the brain stem, where the auditory nerve ends at two collections of neurons called the cochlear nuclei, things start to get complicated. That is as it should be. Very literally, brain processing gets more sophisticated as it ascends, the way a good curriculum builds on itself and asks more of students as they move through school.

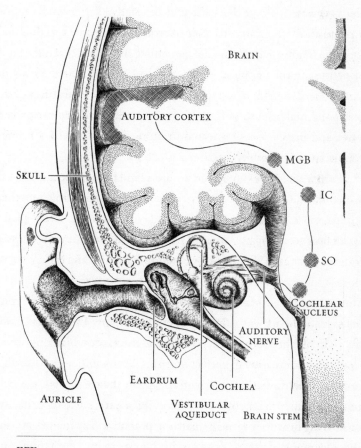

KEY:
SO: Superior olive; IC: Inferior colliculus; MGB: Medial geniculate body

The cochlear nuclei sort the incoming auditory signal along two different tracks, and the various types of cells of the cochlear nuclei each specialize in just one kind of signal. The organization by tone of the basilar membrane, for instance, is repeated in the brain stem, so that some cells respond to high-frequency sounds and others to low-frequency sounds. It is thought that features such as your ability to tell where a sound came from or the fact that you jump at loud noises can be traced to specific cells in the dorsal and ventral pathways respectively. From the cochlear nuclei, sound signals follow these two parallel pathways—along the back and the belly of the brain—in an intricate and complex route passing through regions such as the superior olive (really) and the medial geniculate until they reach the auditory cortex in the temporal lobe, just above the ear where the sound started. Here, too, there are specialized cells to extract features of the sound and make sense of it. With practice, the auditory cortex gets better and better at sophisticated listening, which is why a trained technician, for example, can tune a piano by ear.

If we wanted Alex to be able to listen and talk, he needed first to hear sounds (to recognize simply that someone is speaking) and then to make sense of those sounds (to someday follow a teacher's explanation of how to do long division). The former would require a jet assist from technology. The latter would require practice and time.

Information about plasticity first really reached parents in the mid-1990s, just when I was thinking about having my first child. It was easy to freak out about the responsibilities now incumbent on parents to oversee the creation of perfect brains. The possibilities for learning seemed to be pushed to younger and younger ages. Now it wasn't enough to nurture and care for your child; a parent had to be always considering how to maximize learning potential. Toy manufacturers take advantage of this worry by selling us as many "enriching" and "brain-boosting" products as possible.

Frankly, most middle-class, educated, professional parents like me have the luxury of thinking in terms of enrichment. Our children have the good luck to be born into families where basic needs are easily met; in their early years, these kids are carefully nurtured, fed nutritious meals, talked to, read to, played with. All of that is essential to building strong brain architecture in a way that Baby Mozart tapes and foreign language playgroups are not.

In brain plasticity terms, the flip side of enrichment is deprivation. That's what neurologists call it when a brain lacks stimulation. "If a system can be influenced by the environment and enhanced, that same system is vulnerable because if the right input isn't available then it won't develop optimally," says the University of Oregon's Helen Neville, one of the pioneers in the study of plasticity. You can see this plainly in studies that compare certain brain functions in children from different socioeconomic groups. A study published in 2009 asked both low- and middle-income nine- and ten-year-olds to watch images flashing on a computer. The children were instructed to press a button when a tilted triangle appeared. This is a skill that reflects activity in the prefrontal cortex, which controls planning and executive function. The poorer children were far less able to detect the tilted triangles and block out distractions—so much so that one of the researchers likened the results to what he sees in stroke victims who have lesions in the prefrontal cortex.

I had generally considered all the hype about enrichment just that: hype. Even so, I was susceptible to worry and guilt. I might have forgotten what a cause-and-effect toy was, but like so many of my friends, I had signed my boys up for baby music classes, read to them daily, owned educational puzzles, and so on.

Suddenly, in Alex's case, I was staring deprivation straight in the face. Alex wasn't talking because not enough sound was traveling the pathways that led to the parts of his brain that deal with hearing.

II

WHAT IF THE BLIND COULD SEE?

What exactly did deprivation do to the brain? For a long time, no one really knew and this was mostly a philosophical question, famously encapsulated by Irish scientist and politician William Molyneux, who wrote to John Locke in 1688, What if a blind person who has learned to distinguish a cube from a globe by touch, was suddenly able to see? Would he then be able to recognize these objects by sight? Molyneux and Locke both suspected the answer was no. The question was passed down and contemplated anew through the years like a philosophical folktale. Some philosophers argued yes; others echoed Molyneux and Locke and said no.

As for neuroscientists, once such a job title existed, they thought about such problems in anatomical and physiological terms. What parts of the brain were involved in perception and learning, and how did they work? Well into the twentieth century, it was believed that the brain was unchangeable beyond early childhood. (The few researchers who produced evidence suggesting otherwise were mostly ignored.) Furthermore, it was thought that structure determined function, that each region of the brain could perform only its assigned task. In 1913, Santiago Ramón y Cajal, whom many consider the father of modern neuroscience, wrote that the adult brain's pathways are "fixed, ended, immutable." Even though it was understood that the brains of children were works in progress, the details of how that developmental work got done were mysterious.

A series of experiments with cats in the 1950s and 1960s forced a radical reassessment of accepted ideas about how experience might affect the brain. David Hubel was born and raised in Canada and Torsten Wiesel in Sweden. They met at Johns Hopkins University, where both were conducting research in physiology, and teamed up, ostensibly for a few months; their partnership lasted twenty-five years, most of them spent at Harvard.

Hubel and Wiesel concentrated on vision, beginning with the question: What is it to see? Compared to the machines that let researchers see inside the brain today, they used spectacularly low-tech equipment. Hubel fashioned a lathe himself to make the electrodes they would use to probe the brain cells of cats and monkeys. When they needed a screen on the ceiling on which to project images, they hung white sheets. And they plotted the data they received from their recordings of the electrical activity of individual cells using pencils and paper tacked to the wall.

Their first few years of work, much of it described in a 1962 paper Hubel called their "magnum opus," laid the groundwork for what we know today about how visual information comes into the brain and what happens to it there—cell by cell. But they didn't stop there. Their curiosity had been piqued by the fact that, in children, cataracts led to blindness, whereas in adults, they did not. They were also struck by experiments in which animals raised in darkness were left with impaired vision even though there was nothing wrong with their eyes. So after establishing the physiology of normal vision in adult animals, Hubel and Wiesel asked what happens as vision develops in younger animals and, critically, what happens if there's a problem.

They thought the first logical step was to suture shut one eye of young kittens, so that they could compare the information from each eye. In these kittens, they expected vision to develop relatively normally in the non-deprived eye. When the sutures had been removed and they covered up the good eye, the cats were effectively blind in the eye that had been deprived of vision, even though there was technically nothing wrong with the eye. The damage was obvious. The first cat they experimented with fell off the table when released. "That is something," wrote Hubel, "that no self-respecting cat would do." When they sacrificed the animals and looked at their brains, though, they were surprised to find that the cells in the visual cortex looked radically different from those in cats with normal vision. The good eye had co-opted much of the neurological territory of the deprived eye like a vine that spreads into the neighbor's yard. Was it from a failure to develop or a withering from disuse of connections that were already formed? They weren't sure, but proposed the latter. Hubel and Wiesel also discovered that some of the visual connections wired up in spite of disuse, which meant that some of the neurological functioning had to be innate.

The next round of experiments involved suturing both eyes of kittens. (Science can undoubtedly be cruel.) After about three months, they reopened the cats' eyes. Again, the cats were effectively blind, even though there was nothing wrong with their eyes. The problem lay in their brains, which had no experience processing visual information and were no longer capable of the task. Finally, Hubel and Wiesel wanted to know how variations in the length of deprivation or the moment of its onset made a difference. Cats hardly use their eyes in the first three weeks after birth, but their visual systems are fairly mature by three months of age. It turned out that if kittens were deprived of sight at any point during the fourth through the eighth week of development, the damage was considerable. After three months of age, however, a cat could have its eyes sutured for months with no appreciable consequences.

This seminal work in vision established the idea of "critical periods" in brain development, called sensitive periods today. Synaptic connections in the brain, it seemed, are best created within a certain window of time, but if an area of the brain is deprived of stimulation beyond that sensitive period, it will reorganize and get used for something else.

Naturally, I wondered what happens in deafness. Until the advent of cochlear implants, it wasn't possible to take away and then restore hearing as Hubel and Wiesel did with vision in cats. Even with earplugs, there is some hearing, as anyone who has tried to nap near a construction site knows. Thanks in large part to Hubel and Wiesel, the visual cortex, located at the back of the brain, remained the best understood part of the brain for years—it still is, really. The secrets of the auditory system, which concentrates in the temporal area, have taken more time to unlock.

Helen Neville was one of the first neuroscientists to demonstrate how the brain changes in deafness. Neville grew up in Canada and didn't start out wanting to be a scientist. A product of the 1960s, she wanted to be a revolutionary and bring about social justice. "I was a rabble-rouser. I didn't want to go to university," she told me when we met in her office at the University of Oregon. She was in the middle of finishing several important grant applications, so we were pressed for time, and she spoke in rapid-fire sentences; I suspected she might talk the same way on a lazy Sunday afternoon or over a glass of wine. Her passionate, salty personality—four-letter words and phrases like "crazy-ass religion" were sprinkled through her conversation—was evident. "I just wanted to change the world," she said. She wasn't enough of a rebel to buck her family's insistence that she get an education, however, so she stayed in school. Naturally curious, she took a wide range of courses, including one in something called neurophysiology, an exploration of the workings of the brain. "I found out that ideas were chemical cascades in the brain," she says. "That's when I had this epiphany. *What?!* That's what ideas are? I need to know about that."

After college, she was still a rabble-rouser. "I went to Montreal to try to start the revolution. But, you know, I couldn't start the revolution. Or get a job." When she found out she could get paid to go to graduate school, she did that instead. "I realized that if I really did think that I could change ideas and ultimately the world, then I had to know how changeable the brain was. Ideas are your brain, your mind is your brain. I needed to know as much as I could about the changeability of your mind and your brain." She would bring the revolution to the laboratory.

At the time, "everybody thought the brain was determined and fixed and organized at or before birth," says Neville. Even Hubel and Wiesel, who had shown such dramatic alterations in the brains of kittens, believed that they had found the limits of change. "Of course it wouldn't change with experience" is how Neville recounts the thinking. "How could you know the brain if it was changing every time you turned around?" She put her mind to the problem of how to show that the reigning dogma was wrong. Her solution: to look at the brains of deaf people. Says Neville: "They'd had such extremely different experiences that if their brains weren't different, maybe what all those guys were saying was right."

Although it was popularly imagined that deaf people's sight improved and blind people's hearing was sharper, scientists had been disappointed in their search for evidence of any actual compensatory changes. Studies of the blind had found that they could not necessarily hear softer sounds than sighted people could. And the deaf had not been shown to be any better at perceiving contrast, seeing in dimmer light, or perceiving subtle motion in slow-moving objects. Neville wondered if her predecessors had been looking in the wrong places, measuring the wrong skills. It was 1983 by the time she started on this work. The technological innovations such as fMRI that would transform neuroscience—and Neville's own work—were still another decade away. So she used the most advanced technique available: electroencephalography, EEG, a measure of evoked potentials on the scalp. She calls it "eavesdropping on the brain's electrical conversation." A potential—the metaphorical significance of the term is appealing—is a measurement of the electrical activity in the brain or roughly how many neurons fire and when. An EEG generates a line that moves up and down representing brain wave activity. Electrodes are glued all over a subject's head and scalp (today there are complicated net caps with all the electrodes sewn in to make the job easier), and the subject is asked to look or listen. The waveforms that are generated indicate how quickly the brain responded to whatever the subject heard or saw.

Using a group of subjects who had been deaf from birth, and a group of hearing controls, Neville set up an experiment in which she told the participants to look straight ahead, and then she repeatedly flashed a light off to the side, where it could be seen only in the peripheral vision. Electrical responses to the flashes of light were two to three times higher in deaf brains than in hearing brains. Even more intriguing was the location of the response. It was not primarily over the visual cortex, "which is where any well-behaved brain should be registering flashes of light," wrote Sharon Begley when she described the experiment in her book, *Train Your Mind, Change Your Brain.* Instead, the responses were over the auditory cortex, which conventional wisdom said should be sitting dormant in a deaf person who heard no sound.

Captivated, Neville launched a series of studies trying to pin down what brain functions might have been enhanced in both deaf and blind subjects in examples of what researchers call "compensatory plasticity." Those studies, which some of her postdoctoral fellows have gone on to pursue in their own laboratories, showed that for the deaf, enhancement came in two areas: peripheral vision and motion processing. In one study, for example, Neville had subjects watch a white square on a computer screen and try to detect which way it was moving. If the square was in the central field of vision, there was no difference between the deaf and the hearing. But in the periphery, the deaf were faster and more accurate than the hearing subjects. "All of the early studies had just studied the center," says Neville.

In addition, those previous studies included people who had lost their hearing to meningitis or encephalitis. "They were dealing already with an altered brain," she says dismissively. To avoid confounding variables like that, Neville tested groups of subjects, all of whom had deaf parents. Some subjects were deaf from birth. Others were their hearing siblings, who learned ASL as their first language but had no auditory deprivation. "You want to know what's due to auditory deprivation and what's due to visual and spatial language," she explains. The brains of those hearing native signers did not show the same changes as their deaf brothers' and sisters'. "That's how we sorted it out," says Neville, beaming. "It's beautiful, so beautiful!"

She also wanted to explore in more detail the question of what parts of the brain deaf and blind people were using, a question that got easier to answer with the advent of functional MRI in the early 1990s. While EEGs are useful for indicating the timing of a response, fMRI does a better job of pinpointing its location. MRI, without the "functional" prefix, uses the magnetic field to take pictures inside the body and is used diagnostically by clinicians for all manner of injuries and illnesses. fMRI is different. It produces what is essentially an afterimage of neural activity by showing contrasting areas of blood flow in the brain. By watching how blood moves around the brain—or, more precisely, how the level of oxygen in the blood rises or falls—

researchers can tell which areas of the brain are active during particular activities. When Neville looked at the brains of deaf and blind subjects with fMRI, it suddenly looked very much as if structure did not necessarily determine function. She found that signing deaf adults were using parts of the left temporal lobe, usually given over to spoken language in hearing people, for visual processing. They were still using that area of the brain for language, but it was a manual, visual language. In blind subjects, the reverse was true: They had enhanced auditory attention processing in certain areas of the visual cortex.

For a signing deaf person, capitalizing on unused portions of the brain makes sense. It seems likely that we start out with redundant connections between auditory and visual areas, and that they can be tweaked by experience, but that the overlap gradually decreases. If you want to use a cochlear implant, however, you need to maximize the auditory cortex's original purpose: hearing. What were the critical periods for the auditory cortex? After what point was it forever reorganized? Or was it? These were the questions Helen Neville proposed and other researchers set out to answer in the late 1990s, once they had enough subjects with cochlear implants to study the issue.

In the Department of Speech, Language and Hearing Sciences at the University of Colorado at Boulder, auditory neuroscientist Anu Sharma studies brain development in children with cochlear implants. It's work she began at Arizona State University with speech and hearing scientist Michael Dorman, who started working on cochlear implants in the 1980s. In order to assess how well a brain is making use of the sensory input it receives, Sharma focuses on how long it takes the brain to react to sound. The speed of that reaction is a measurement of synaptic development. Sharma, too, uses EEG, looking particularly at what is known as the cortical auditory evoked potential (CAEP for short), the response to sound beyond the brain stem in the auditory cortex.

Studying the waveform that is generated by the test, Sharma looks for the first big fall and rise of the line: The first valley is known as P1 (first positivity), and the adjacent peak is N1 (first negativity). The upward slope of the line connecting those two points roughly indicates the speed at which two parts of the brain—primary cortex and higher order cortex—are communicating. Sitting in her office in Boulder with a view of the Rocky Mountains behind her, Sharma demonstrates by holding out her fist to represent the primary cortex, then wrapping her other hand over the top of her fist to show the higher order cortex (the outer layer of the brain). "They need to connect; they need to talk," she says.

In typically developing hearing children, that connection in the brain starts slowly and then speeds up. Latency is the time it takes in milliseconds for the brain to react to a sound. In a newborn the latency of the P1 response might be three hundred milliseconds; in a three-year-old, it has dropped to one hundred twenty-five milliseconds. As a child grows, that time continues to shorten. With age and experience, the brain becomes more efficient. By adulthood, P1 latency is down to sixty milliseconds.

In children with cochlear implants, Sharma found that being implanted early makes all the difference in terms of how much help the child will get from the implant. The peak of synaptic activity in the auditory cortex is when a child is three and a half, says Sharma. Experience prunes and refines synapses allowing learning to occur. Those that are getting used stay; those that are not are pruned away. "Hearing helps to prune them," she explains. "If a child gets the implant under the age of three and a half, that part of the brain looks quite similar to that of a normal hearing child. If they wait until they are seven or older [or had been deaf for more than seven years], the hearing part of the brain of the implanted child never looks the same." Between the ages of three and seven, the studies showed a range of responses, but earlier was almost always better. Sharma and her colleagues had identified sensitive periods for hearing and brain development.

"We looked more closely and found that what happens after the sensitive period closes is the brain gets reorganized," she says. "That's important real estate. If sound is not going in, it's not going to sit there waiting forever." In some of the secondary areas, for instance, vision and touch take over. "That's how the brain changes in deafness."

No wonder Dr. Parisier was in a hurry.

12

CRITICAL BANDWIDTHS

On a sunny spring day in 1966, Graeme Clark, a young Australian ear, nose, and throat surgeon, had a little extra time for lunch and decided to eat outside on a park bench. He was shuttling between jobs at various hospitals in Melbourne, so he carried a lot with him, including a backlog of scientific and medical journals. He pulled one out and found a report by Blair Simmons on achieving some hearing sensation, though no speech, from the six-channel device they had implanted in Anthony Vierra. Just like that, Clark knew what he was going to do with the rest of his life. "That lit that research fire in the belly," he says. "It all became clear."

We were sitting around a table at the Royal Victorian Eye and Ear Hospital in Melbourne. For some thirty-five years until his retirement in 2004, this had been Clark's office. It was here, about as far from the medical centers of Europe and America as it is physically possible to be, that Clark assembled a team of young researchers to build a viable multichannel cochlear implant. Research groups at the University of California, San Francisco (UCSF), and at the University of Utah as well as in Paris and Vienna were pursuing the same goal. (Bill House was still using his single-channel device and Blair Simmons, who died in

1998, got less involved over time.) At the start, the idea that Clark's group could achieve such an ambitious goal was almost laughable. They had no money, little prestige, and, at times, no patients. Today, many of those same audiologists, engineers, and scientists are still in Melbourne, running the various spin-off clinical and research organizations that grew out of Clark's efforts.

A thoughtful, deeply religious man, Clark shares the formality of his generation. In a country where men have been wearing shorts to work for decades, he is often in a jacket and tie, as he was when we met at the hospital. Earlier in his career, at a dinner in his honor, a colleague joked that he had wanted to entertain the crowd with tales of Clark's youthful indiscretions but hadn't been able to find any. From a young age, Graeme Clark was apparently remarkably single-minded.

He was born in 1935 in Camden, then a small country town some forty miles from Sydney. His father, Colin, was a pharmacist and ran a chemist shop in town. Around the age of twenty, Colin Clark began to lose his hearing, and it got progressively worse throughout the rest of his life. There could be uncomfortable consequences from the combination of hearing loss and a pharmacy. "People had to speak up to say what they wanted," remembers Graeme Clark. "It was embarrassing when someone came in to ask for women's products or men's products, contraceptives or the like." As a boy of about ten helping out in the shop, Graeme would have to find the requested item. "I didn't know much about the facts of life, but I knew where [things were in the shop]," he says with a laugh. Later in life, Clark asked his father about his experience of hearing loss. "He said it was terrible," Clark remembers. "It was a struggle to hear people at work and social environments with background noise. He would be tired out through having to listen so hard and make conversation." Clark's childhood sense of lost opportunity was strong. "[My father] was quite a sociable person. He could have been president of the [local social] club, but he couldn't manage those things. He'd sit in the corner sometimes of the lounge room when we'd have visitors, and they would think he was dumb."

Life in his father's chemist shop exposed Clark not just to deafness but also to medicine. He got to know the local doctors, and the idea of helping people appealed to him. As early as kindergarten, he told his teacher he wanted "to fix ears" when he grew up. He repeated that to the family's minister when he was ten. By sixteen, he was studying medicine at the University of Sydney. By twenty-nine, he was a consultant surgeon, still so boyish in appearance a nurse once refused to let him in to visit his own patient.

But medicine alone wasn't enough. Clark always nursed an interest in research, believing that a man with both clinical and laboratory experience could be most effective and most aware of what was needed. Blair Simmons's report gave Clark the research direction he'd been missing, one that aligned perfectly with his childhood mission of "fixing ears." He knew Bill House had already operated on a few patients as well, but Clark hewed closer to Simmons's methodical approach. In order to see "what the science would show," he left his flourishing surgical practice to pursue a PhD in auditory physiology, studying electrical stimulation in the auditory system of cats. "It was a complete gamble," he says. "Ninety-nine percent of people said it wouldn't work." Clark and his wife, Margaret, had two children at the time and would go on to have three more. As a doctoral student, he made so much less than he had as a surgeon that when their old car broke down, they couldn't replace it. "But those were some of the happiest times in our lives because I wasn't in the rat race," he remembers. "And because research was exciting."

With his doctorate in hand, Clark got lucky. A position as chair of the Department of Otolaryngology opened up at the University of Melbourne and the Royal Victorian Eye and Ear Hospital. As Clark worked to assemble a team there, he was raising money with one hand—literally shaking a can on street corners at times as part of fund-raising efforts—and doling out the cash carefully to his young staff with the other. Everyone worked from one three-month contract to the next. "I was young and they were younger," he says. Like Simmons, Clark believed he would need a multidisciplinary team to provide expertise on the considerable variety of jobs required to create this piece of technology. Over the years, he employed engineers, animal researchers, audiologists, computer programmers, speech scientists, and surgeons.

For most of the 1970s, Clark's group labored over a workable prototype of a multichannel device he called a "bionic ear." The solution to one particularly stubborn surgical question—how to insert the device safely and fully into the snail-shaped cochlea—came to Clark during a vacation. While his children played on the beach, he collected a series of spiral-shaped shells and a variety of grasses and twigs, and tried to stuff the grasses through the shells. He discovered that materials that were stiff at one end and flexible at the other worked perfectly. In the lab, his team created an electrode bundle that mimicked the grasses he'd found—stiff at the base but increasingly flexible toward the tip. Other challenges included reducing the necessary circuits from an original diagram wider and higher than a grown man's torso to a tiny silicon chip.

The resulting device looked as homemade as it was. Bigger, lumpier, and less refined than the corporate versions that followed, the implanted piece consisted of a gold box a few centimeters square containing the electronics to receive signals and stimulate the internal electrodes. The electrodes were attached to the stimulator by a connector so that in the event of a failure, only the gold box would have to be replaced, not the electrodes in the inner ear. The whole package was encased in silicone to protect against corrosive bodily fluids.

But they still had to figure out what signals to send through this new device. "The way sound stimulates the inner ear is different from the way in which electrical currents stimulate the nerves," explains

Clark. "When you put electrical current into the nerves, it tends to stimulate them all at one time." The answers wouldn't really come until after they had implanted their first patient.

That was another problem. They needed patients, and they didn't have any. Because of the fund-raising campaign, which included tele-thons, the public profile of their work was high. But doctors who saw patients refused to refer anyone. They thought Clark had overstated what was achievable. "I think I said there would be about five thou-sand people in Australia who could benefit from a cochlear implant," says Clark with a laugh. In fact, he understated the eventual demand by several orders of magnitude—an estimated 320,000 people use them worldwide today. But at the time, in Australia as in the United States, the response in medical and scientific circles was skeptical.

Clark was getting desperate until he met a forty-eight-year-old man named Rod Saunders who had been visiting the deafness unit in Clark's own hospital. Saunders had lost his hearing in a car accident twelve months earlier, when lumber he was carrying in his car—precariously jammed between the seats—smashed into his head when he collided with a light pole. His injuries left him completely deaf in both ears. He had seen Clark's project in the news. When he came in for an appointment one day, he asked at the front desk whether he could see Professor Clark. "No," he was told. "We don't recommend it." Standing within earshot, however, was an audiologist named Angela Marshall, who was spending some of her time in Clark's laboratory. She pulled Saunders and his wife aside and said she could help. As Clark puts it, "Angela smuggled Rod up."

The second patient, George Watson, managed to get to Clark directly. He was a World War II veteran and had been profoundly deaf for thirteen years after losing his hearing progressively following a bomb blast. Both Saunders and Watson felt they had little to lose, and both found their deafness debilitating. Saunders, who couldn't speech-read well at all, called it a "nightmare." He told Clark what he missed most was hearing the voices of his family. "I miss music," he said. "I even miss the sound of the dog barking." Watson described his feeling of isolation. "You feel completely alone," he said. "I mean you go to a football match, people cheering and so forth, or you go to a race meeting, but it is all so very even all the time. . . . Actually, it is very boring. Everything is the same, nothing seems to alter." Even these conversations with Clark were awkward, requiring a combination of speechreading and written questions.

"We were really fortunate with the first two guys," says Richard Dowell, who today heads the University of Melbourne's Department of Audiology and Speech Pathology but in the late 1970s was a twenty-two-year-old audiologist working for Clark on perception testing with Saunders and Watson. "They were both sort of laconic characters, not too fussed about anything. They had to go through a lot of boring stuff and things didn't [always] go right. They were basically guys who'd put up with anything and continue to keep coming in and support the work. They didn't necessarily want anything out of it for themselves."

Clark decided to implant Rod Saunders first and scheduled surgery for August 1, 1978. He and the surgeon who would assist him, Brian Pyman, had repeatedly rehearsed the steps of the operation on human temporal bones from the morgue. When he could practice no more, Clark went away with Margaret for a prayer weekend. The operation lasted more than eight hours. As Saunders recovered that night in the ward, Clark called the night nurse every few hours to check on his condition, but all was well. Saunders went home after a week.

Three weeks later—enough time for the surgical wound to heal—Saunders returned to the hospital to have the prosthesis turned on. They put the external coil in position and turned on the electrical current, gradually increasing its strength.

It didn't work.

"Rod, do you hear any sound?" Clark asked.

Saunders was listening intently, but he answered dejectedly. "I'm sorry. I can only hear the hissing noises in my head."

Depressed and worried, Clark and his colleagues had to send Saunders home. When he returned a few days later, results were no better. There were several sleepless nights for Clark. Finally, before Saunders's third appointment, engineer Jim Patrick discovered a fault in the test equipment and repaired it—to everyone's relief. "We approached the next session with great anticipation," remembered Clark.

This time Saunders heard sounds. Each of the electrodes was tested and they were all working. The sounds were limited, but a cause for celebration nonetheless. Clark and Pyman took the surgical nurses and their spouses out for a Chinese dinner to mark the occasion.

At the next session, they wanted to know whether Saunders could recognize voicing and the rhythm of speech. They used the computer to play songs through the implant, beginning with the Australian national anthem, "God Save the Queen." Immediately, Saunders stood to attention, pulling out the wires connecting him to the computer when he did. Everyone laughed with relief. Then someone suggested they try "Waltzing Matilda." Saunders recognized that as well. It was solid progress, but Saunders still couldn't understand speech and he didn't seem to be recognizing different pitches. The design of this device was predicated on the idea that multiple electrodes laid out along the cochlea would deliver variations in frequency that would enable users to hear the sounds of words. But Saunders described the different signals he was hearing as "sharp" at high frequencies and "dull" at the low end, but not higher or lower than one another. "The sensations were changing in timbre, not pitch," explains Clark. He couldn't help but wonder: "Have we gone to all the trouble to produce a multichannel system . . . and it didn't work?"

An engineer named Jo Tong was one of Clark's closest collaborators in those early days, and Tong had taken the lead in designing the speech processing program that determined the instructions sent into the new device. He and Clark began with the belief that the cochlear implant had to try to reproduce nearly everything that goes on in a normal cochlea. Like a glass prism breaking up light into all the colors of the rainbow, the cochlea takes a complicated sound such as speech and breaks it into its component frequencies. The internal electrodes of the implant were designed to run along the cochlea, mimicking the natural sequence of frequencies. Clark's team was betting that the *place* that was stimulated was critical to delivering pitch to the user, and pitch was critical to understanding speech. Initially, Tong designed a processing program that stimulated every electrode, on the theory that all parts of your basilar membrane are perceiving sound at once in normal hearing and that the appropriate areas would react more strongly to the appropriate frequencies. But since electrical stimulation is far less subtle than the workings of a normal cochlea, the result had proved incomprehensible.

So Tong and Clark hit on plan B, which was to try to extract the elements that convey the most information in speech and send only those through the implant. The new version was almost as simple as the first one had been complex. It made use of formants, the bands of dominant energy first described at Bell Labs that vary from one sound to another. If we produced sounds only with the larynx, we wouldn't get anything but low frequencies. However, those low frequencies contain harmonics up to ten or twenty times higher. If the larynx is vibrating a hundred times per second—at 100 Hz—the harmonics are at 200 and 300 Hz, up to 2,000 Hz or higher. "As you move your lips and tongue and open and close the flap that's behind your nose to create speech sounds, all those things modify the amplitude of all the different harmonics," says Hugh McDermott, an acoustic engineer who joined Clark's team in 1990. For each sound, the first region where the frequencies are emphasized—where the energy is strongest—is called the first formant, the next one going up is the second formant, and so on. The new speech processing program was known as F0F2 (or "F naught, F two" when Australians speak of it). That meant that the program extracted only two pieces of information from each speech sound: its fundamental frequency (F0) and the second formant (F2). "It's the first two formants that contain nearly all of the information that you need to understand speech," says McDermott. The second formant is also difficult to see on the lips, so it was a particularly useful extra piece of information. "If you have to nail down just one parameter," says McDermott, "that's the one to choose."

F0F2, in other words, was lean and mean. Recognizing speech on the basis of a few formants is like identifying an entire mountain range from the outline of only its two most distinctive peaks. It worked by having one electrode present a rate of electric pulses that matched the vibration—the fundamental frequency—of the larynx. Then the second formant, which might be as high as 1.5 kHz (kilohertz), was represented on a second electrode. That second formant moved around from electrode to electrode according to the speech sounds created. F0F2 sounded even more mechanical and synthetic than implants do today, and it took a lot of getting used to, but it worked because only one electrode was on at a time, eliminating the problem of overstimulation.

With this new processing system in place, Saunders began to understand some limited speech. He could be 60 to 70 percent correct on tests that asked him to identify the vowels embedded in words: "heed," "hard," "hood," "had," and so on. As the end of 1978 neared and money was running short again, Clark insisted they try Saunders on a harder test: what's known as open-set speech recognition. Until then, they had done only closed sets—reciting words that were part of familiar categories, such as types of fruit. Speech in real life, of course, isn't so predictable; open-set testing throws wide the possibilities. Angela Marshall was hesitant, fearing it wouldn't work. "I said, if we fail, we fail," says Clark. As the group stood watching with bated breath, Marshall presented one unrelated word at a time.

"Ship," she said.

"Chat," replied Saunders. Completely wrong.

"Goat."

"Boat," said Saunders. Closer.

"Rich."

"Rich," said Saunders. He had gotten one right!

By the end of the tests, Saunders had gotten 10 to 20 percent of the open-set words correct. That's not the least bit impressive by today's standards, but it was hugely significant at the time for a man who was profoundly deaf. Clark was overcome. "I knew that had really shown that this was effective," he says. He pointed down the hallway of the hospital where we sat. "I was so moved, I simply went into the lab there and burst into tears of joy." The only other time he cried in his adult life, he told me, was when he and Margaret worried over the health of one of their five children.

George Watson became the second Australian to be implanted, in July 1979. His early audiological results were promising enough that Lois Martin, who had taken over for Angela Marshall, "decided to go for broke," says Clark, and read Watson some lines from the daily newspaper, then asked him to repeat what she'd said. "He'd nearly got it all right," says Clark, who was sitting in his office up the hall working at the time. "There was great excitement and they came rushing up the corridor to tell me," he remembers. They had wondered what Watson's brain would remember of sound. "George was showing us that he could remember the actual sound that he'd heard thirteen years before." Clark had only two patients at this point, "but the two of them together told us a lot about what was possible."

Not that there weren't some disasters. Until that time, both Saunders and Watson were only able to use their implants in the laboratory, hooked up to the computer. Clark instructed an engineer named

Peter Seligman to develop a portable device that they could take home. "It was as big as a handbag," remembers Dowell. "We thought it was fantastic." They called a press conference to announce that two patients were successfully using a portable cochlear implant.

"That day, George's implant failed," says Dowell. "I was preparing him to go out there to talk to the press, and he told me that it had stopped working." Watson reckoned he could wing it using his speechreading skills. And he did. Watson told the assembled reporters how wonderful his implant was. "None of those guys who were there that day would have had a clue that he wasn't actually hearing anything," says Dowell. Rod Saunders, on the other hand, although his implant was working, appeared to be struggling more because he had such a hard time reading lips. "It was reported as a great break-through," Dowell says, laughing. "It's so long ago I can tell the story." Clark, too, is willing to tell the story today. "If they'd asked, I would have to have said the implants failed. I just sort of held my breath and they didn't ask."

Meanwhile, in San Francisco, another team had been pursuing a similar path. Auditory neuroscientist Michael Merzenich had been recruited to the University of California, San Francisco, in part because a doctor there, Robin Michelson, was pursuing a cochlear implant. Two of Michelson's patients participated in the review performed at the University of Pittsburgh by Robert Bilger. The head of UCSF's otolaryngology department, Francis Sooy, thought the idea of cochlear implants had merit, but that it needed a more scientific approach.

"When I showed up at UCSF, I was intrigued by the idea, but when I talked to Michelson, I realized that he understood almost nothing about the inner ear and nothing about auditory coding issues," Merzenich told me. "He had the idea if we just inject the sound, it will sort it out. He talked about what he was seeing in his patients in a very grandiose way." As an expert in neurophysiology, Merzenich found such talk hard to take, especially since he was engaged in a series of studies that revealed new details about the workings of the central auditory system. He discovered that information wasn't just passed along from level to level—from the brain stem to the cochlear nuclei to the superior olive and so on. Along the way, the system both pulled out and put back information, always keeping it sorted by frequency but otherwise dispersing information broadly. "Our system extracted information in a dozen ways; combined it all, then extracted it again; then combined and extracted again; then again—just to get [to the primary auditory cortex]," Merzenich later wrote. "Even this country boy could understand the potential combinative selectivity and power of such an information processing strategy!" On the other hand, he could also see how hard it would be to replicate.

Robin Michelson, it must be said, was a smart man. He originally trained as a physicist, two of his children became physicists, and his uncle Albert Michelson won the Nobel Prize in Physics. Several people described him to me as a passionate tinkerer, with grand ideas but not always the ability to see them through. Merzenich couldn't really see how such a device as Michelson described would ever be usable clinically as anything more than a crude aid to lipreading. Unresolved safety issues worried him as well. "The inner ear is a very fragile organ," he says, "and I thought that surely introducing these electrodes directly into the inner ear must be pretty devastating to it and must carry significant risk."

Besides, Merzenich was soon busy conducting what would turn out to be revolutionary studies on brain plasticity. That was one of the reasons I had been so eager to talk to him. For someone like me, who wanted to know about both cochlear implants and how the brain changed with experience, there was no other researcher in the world who had played such a major role in both arenas. By the time I wrote to him, Merzenich was serving as chief scientific officer (and cofounder) at a scientific learning development company called Posit

Science, whose flagship product is BrainHQ, a brain training plat-form. We met in their offices in downtown San Francisco. "If you think about it," he told me, "[the cochlear implant] is the grandest brain plasticity experiment that you can imagine."

As a postdoctoral fellow at the University of Wisconsin, Mer-zenich had participated in an intriguing study showing changes in the cortex of macaque monkeys after nerves in their hands were surgi-cally repaired. He hadn't quite tumbled to the significance of these findings, but shortly after he arrived at UCSF in 1970, he decided to pursue this line of research as well as his auditory work. Together with Jon Kaas, a friend and colleague at Vanderbilt University, Mer-zenich set up a study with adult owl monkeys. By laboriously touching different parts of the hand and recording where each touch registered in the brain, they created a picture that correlated what happened on the hand to what happened in the brain. With the "before" maps com-plete, the researchers severed the medial nerve in the monkeys' right hands, leaving the animals unable to feel anything near the thumb and nearby fingers. (Here, too, the requirements of science are discomfit-ing, to say the least.) According to standard thinking on the brain at the time, that should have left a dead spot in the area that had previ-ously been receiving messages from the nerve. A few months later, after the monkeys had lived with their new condition for a time, Mer-zenich and Kaas studied the animals' brains. Completely contrary to the dogma of the day, they found that those areas of the brain were not dead at all. Instead, they were alive with signals from other parts of the hand.

In another study, one of Merzenich's collaborators, William Jen-kins, taught owl monkeys to reach through their cages and run their fingers along a grooved spinning disk with just the right amount of pressure to keep the disk spinning and their fingers gliding along it. When Jenkins and Merzenich studied the maps of those monkeys' brains, they found that even a task such as running a finger along a disk—much less dramatic than a severed nerve—had altered the map. The brain area responding to that particular finger was now four times larger than it had been.

Finally, Merzenich understood what he was seeing: that the brain is capable of changing with experience throughout life. It was a finding that changed his career. It also brought considerable controversy. "Initially, the mainstream saw the arguments we made for adult plasticity in monkeys as almost certainly wrong," he recalls. Major scientists like Torsten Wiesel, whose Nobel Prize had been awarded in part for establishing the critical period, was one of those who said outright that the findings could not be true. It was years before the work was widely accepted.

Right away, however, the work on plasticity prodded Merzenich's thinking about the development of cochlear implants. Robin Michelson had been persistent. "For a year, this very nice man pestered me and pestered me and pestered me," remembers Merzenich. "He would come to see me at least once or twice a week and tell me more about the wondrous things he'd seen in his patients and bug me to come help him." For a year, Merzenich resisted. "Finally, in part to get him off my back, I said okay." Merzenich agreed to do some basic psychophysical experiments—create some stimuli and apply them—and see what Michelson's best patient could really hear.

The patient was a woman named Ellen Bories from California's Central Valley. "A great lady," says Merzenich fondly. "We started trying to define what she could and couldn't hear with her cochlear implant, and I was blown away by what she could hear with this bullshit device." Michelson's implant, at that point, was "a railroad track electrode," as Merzenich had taken to calling it dismissively, that delivered only a single channel of information. While Bories could not recognize speech, she could do some impressive things. In the low-frequency range, up to about 1,000 Hz, she could distinguish between the pitches of different vowels—she could hear and repeat "ah, ee, ii, oh, uh." She could tell whether a speaking voice was male or female. When they played her some music, says Merzenich, "she could actually identify whether it was an oboe or a bassoon. I was just amazed what she could get with one channel."

Not surprisingly, he now saw the effort in a whole new light. For one thing, "I realized how much it could mean even to have these simple, relatively crude devices," he says. "She found it very helpful for lipreading and basic communication and was tickled pink about having this." Secondly, and significantly, Merzenich saw a way forward. At Bell Labs, researchers had never stopped working on the combination of sound and electricity. In the 1960s, they had explored the limits of minimizing the acoustic signal. "They took the complex signal of voice and reduced it and reduced it and reduced it," explains Merzenich. "They said, How simply can I represent the different sound components and still represent speech as intelligible?"

Even earlier, in the 1940s, Harvey Fletcher had introduced the concept of critical bandwidth, a way of describing and putting boundaries on the natural filtering performed by the cochlea. Although there are thousands of frequencies, if those that are close together sound at the same time, one tone can "mask" the other or interfere with its perception, like a blind spot in a rearview mirror. Roughly speaking, critical bandwidth defined the areas within which such masking occurred. How many separate critical bandwidths were required for intelligible speech? The answer was eleven, and Merzenich realized it could apply to cochlear implants, too. The implanted electrodes would be exciting individual sites, each of which represented specific sounds. "The question was how many sites would I need to excite to make that speech perfectly intelligible. Eleven is sort of the magical number."

He was able to simplify this a little further, because Bell Labs had also built a voice coder (or vocoder) in which you allowed the base band, the lowest-frequency bandwidth, to do more than its fair share of the work and represent information up to about 1,000 Hz. With this one wider channel doing the heavy lifting on the bottom, Merzenich calculated that he needed a minimum of five narrower bands above it, for a total of six channels. Modern cochlear implants still use only up to twenty-two channels—really a child's toy piano compared to natural hearing, which has the range of a Steinway grand. As Merzenich aptly puts it, listening through a cochlear implant is "like playing Chopin with your fist." It was not going to match the elegance that comes from natural hearing; "it's another class of thing," he says. But on the other hand, he was beginning to understand that perhaps it didn't need to be the same.

It didn't need to be the same, that was, if the adult brain was capable of change. Merzenich's cochlear implant work had begun to synthesize with his studies of brain plasticity. "It occurred to me about this time, when we began to see these patients that seemed to understand all kinds of stuff in their cochlear implant, that the recovery in speech understanding was a real challenge to how we thought about how the brain represented information to start with," he says.

Unlike Merzenich, Donald Eddington was far from an established scientist when he got involved with cochlear implants at the University of Utah. As an undergraduate studying electrical engineering, he had a job taking care of the monkey colony in an artificial organ laboratory founded by pioneering researchers William Dobelle and Willem Kolff. Since he was hanging around the lab, Eddington began to help with computer programming and decided to do his graduate work there. Dobelle and Kolff were collaborating with the House Ear Institute in Los Angeles. While Bill House was continuing to work on single-channel implants, another Institute surgeon, Derald Brackmann, wanted to develop multichannel cochlear implants. So that's what Don Eddington did for his PhD dissertation.

On the question of how to arrange the three basic elements—microphone, electronics, and electrodes—needed to make an implant work like an artificial cochlea, the groups in Australia and San Francisco had arrived at similar solutions. They implanted the electrodes, kept the microphone outside the head, and divided the necessary electronics in two. The external part of the package included the speech processor. Internally, there was a small electronic unit that served as a relay station, receiving the signal from outside by radio transmission through the skin and then generating an electrical signal to pass along the electrodes to the auditory nerve.

In Utah, they did it differently. Eddington's device, ultimately known as the Ineraid, implanted only the electrodes and brought them out to a dime-size plug behind the ear, just as Bill House's prototypes had done. The downside to this approach was that most people found the visible plug rather Frankenstein-like, and the subjects who worked with Eddington could hear only when they were plugged into the computer in the laboratory. From a scientific point of view, however, the strategy had much to recommend it, because it gave the researchers complete control and flexibility. "The wire lead came out and [attached] to a connector that poked through the skin," says Eddington, who moved to Massachusetts Institute of Technology in the 1980s. "You could measure any of the electrode characteristics, and you could send any signal you wanted to the electrode." Rather than extract specific cues from the speech in the processor, as the Australians had, Eddington and his team split the speech signal according to frequency and divided it—all of it—across six channels arranged from low to high.

Over the next dozen or so years, these various researchers— sometimes in collaboration but mostly in competition—worked to overcome all the fundamental issues of electronics, biocompatibility, tissue tolerance, and speech processing that had to be solved before a cochlear implant could reliably go on the market. "To make a long story short, we solved all of these problems," Merzenich says. "We resolved issues of safety, we saw that we could implant things, we resolved issues of how we could excite the inner ear locally, we created electrode arrays and so forth. Pretty much in parallel, we all created these models by which we could control stimulation in patterned forms that would lead to the representation of speech."

As researchers began to see success toward the end of the 1970s and into the 1980s, Merzenich was struck by the fact that initially at least, the coding strategies for each group were very different. "There was no way in hell you could say that they were delivering information in the same form to the brain," he says. "And yet, people basically were resolving it. How the hell can you account for that? I began to understand that the brain was in there, that there was a miracle in play here. Cochlear implants weren't working so well because the engineering was so fabulous, although it's good—it's still the most impressive device implanted in humans in an engineering sense probably. But fundamentally, it worked so marvelously because God or Mother Nature did it, the brain did it, not because the engineers did it."

Of course, the engineers like to take credit. "Success has many fathers," Merzenich jokes. "Failure is an orphan." Indeed, there are quite a few people who can claim—and do—the "invention" of the cochlear implant. The group in Vienna, led by the husband and wife team of Ingeborg and Erwin Hochmair, successfully implanted the first multichannel implant the year before Graeme Clark, although they then pursued a single-channel device for a time. Most agree the modern cochlear implant represents a joint effort.

In the 1980s, corporations got involved, providing necessary cash and the ability to manufacture workable devices at scale. After the Food and Drug Administration approved Bill House's single electrode device for use in adults in 1984, the Australian cochlear implant, by then manufactured by Cochlear Corporation, followed a year later, again with FDA approval only for adults. The San Francisco device was sold to Advanced Bionics, and went on the market a few years

13

SURGERY

In the predawn darkness of a December morning, I watched Alex sleeping in his crib for a moment and gently ran my fingers along the side of his sweet head just above his ear. In a few hours, that spot would be forever changed by a piece of hardware. Glancing out the window, I saw that Mark had the car waiting, so I scooped Alex up, wrapped him in a blanket against the cold, and carried him outside.

The testing was done. Our decision had been made and it had not been a difficult one for us. As I once told a cochlear implant surgeon, "You had me at hello." Everything about Alex suggested he had a good chance at success. The surgery had been scheduled quickly, just a week or so after our meeting with Dr. Parisier. There was nothing left to do but drive to the hospital down the hushed city streets, where the holiday lights and a blow-up snowman on the corner struck an incongruously cheerful note. Alex pointed to the snowman and smiled sleepily.

Although I knew the risks, surgery didn't scare me unduly. In my family, doctors had always been the good guys. After a traumatic brain injury, my brother's life was saved by a very talented neurosurgeon.

Another neurosurgeon had released my mother from years of pain caused by a facial nerve disorder called trigeminal neuralgia. There'd been other incidents—serious and less serious—all of them adding up to a view of medicine as a force for good.

That didn't make it easy to surrender my child to the team of green-gowned strangers waiting under the cold fluorescent light of the operating room. I was allowed to hold him until he was unconscious. As if he could make himself disappear, Alex buried his small face in my chest and clung to me like one of those clip-on koala toys I had as a child. I had to pry him off and cradle him so that they could put the anesthesia mask on him. When he was out, I was escorted back to where Mark waited in an anteroom. For a time, the two of us sat there, dressed in surgical gowns, booties, and caps, tensely holding hands.

"You'll have to wait upstairs," said a nurse finally, kindly but firmly. "It'll be a few hours."

Though I wasn't there, I know generally how the surgery went. After a little bit of Alex's brown hair was shaved away, Dr. Parisier made an incision like an upside-down question mark around his right ear, then peeled back the skin. He drilled a hole in the mastoid bone to make a seat for the receiver/stimulator (the internal electronics of the implant) and to be able to reach the inner ear. We had chosen the Australian company, Cochlear, that grew out of Graeme Clark's work. Parisier threaded the electrodes of Cochlear's Nucleus Freedom device into Alex's cochlea and tested them during surgery to be sure they were working. So as to avoid an extra surgery, Parisier also replaced the tubes that Alex had received nearly a year earlier, which added some time to the operation. Then he sutured Alex's skin back over the mastoid bone, leaving the implant in place.

The internal implant looks surprisingly simple. But, of course, the simpler it is, the less can go wrong. It has no exposed hard edges. There's a flat round magnet that looks like a lithium battery in a

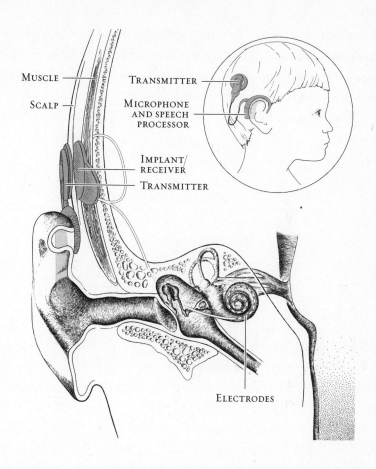

MUSCLE

SCALP

TRANSMITTER

MICROPHONE
AND SPEECH
PROCESSOR

IMPLANT/
RECEIVER

TRANSMITTER

ELECTRODES

camera. It is laminated between two round pieces of silicone and attached to the receiver/stimulator, the size of a dollar coin. Then a pair of wires banded with electrodes extends from the plastic circle like a thin, wispy, curly tail.

More than four hours after the surgery began, Alex was returned to us in the recovery room with a big white bandage blooming from the right side of his head. As he had been all through the previous year of testing and travail, he was a trooper. He didn't cry much. He didn't even get sick from the anesthesia, as if somehow he had decided to do this with as little fuss as possible. When he was fully awake, we took him upstairs to the bed he'd been assigned, where I could lie down with him. It was a few more hours before they let us go home, but finally I wrapped him in his blanket and carried him back into the car, which once again Mark had waiting at the curb.

It was done.

PART TWO

SOUND

14

FLIPPING THE SWITCH

W hat did you do today?" I asked.

It was late on a March afternoon, one month before Alex's third birthday. Matthew, Alex, and I were sitting at the table having a snack. Immediately, Matthew launched into a detailed recitation of a day in the life of a four-year-old preschooler.

"We played tag in the yard and I was it and I chased Miles and we had Goldfish for snack and I made a letter 'B' with beans and . . ." Matthew loved to talk. Words poured out of him like water. When he stopped for a breath, I turned to Alex, sitting to my left.

"Alex, what about you?"

I didn't expect an answer, but I always included him. This time, he surprised me.

"Pay car," he said.

"Pay car. Pay car," I murmured to myself, trying to figure out what he was saying. "You played cars?"

"Pay car," he repeated and added, "fren."

"With friends?" I asked. "Which friends?"

"Ma. A-den."

"Max? Aidan? You played cars with Max and Aidan," I exclaimed. "That's great."

Then Alex began to sing.

"Sun, Sun . . . down me."

As he sang, he held his hands to form a circle and arced it over his head like the sun moving across the sky.

It was a very shaky but recognizable version of the "Mr. Sun" song, complete with hand movements, which his teachers often sang on rainy days.

I sang it back to him.

"Mr. Sun, Sun, Mr. Golden Sun, please shine down on me!"

His face brightened as he made the sun rise and set one more time.

We're talking! I realized. Alex and I are having a conversation!

Two months earlier, in the middle of January, four weeks after Alex's surgery—enough time for the swelling to go down and his wound to heal—we had visited the audiologist to have his implant turned on, a process called activation. It means adding external components and flipping the switch.

Although the entire device is called a cochlear implant, only part of it is actually implanted, the internal piece that receives instructions from the processor and stimulates the electrodes that surgeons place in the inner ear. On its own, that piece does nothing. Like a dormant seed that needs water and sunlight to germinate, it waits for a signal. Producing that signal is the job of the external piece, which contains the microphones, battery pack, and most of the electronics. This outer piece has evolved from a bulky body-worn case to—in its smallest iteration—a behind-the-ear device not much bigger than a hearing aid. The intention was always to keep the internal components as simple as possible so that technology upgrades could be kept to the outside.

Alex and I sat down with the audiologists we'd been assigned, Lisa Goldin and Sabrina Vitulano. They had a big box from Cochlear waiting for us with the external parts of the implant nestled inside in thick black protective foam. Because Alex was young and his ears were small, we were going to start by keeping the battery pack and processor in a separate unit, about the size of a small cigarette case, in a cloth pocket that pinned to his shirt. A cable would run from there to the much smaller piece on his ear, and a second cable would lead to the plastic coil and magnet on his head. When he got bigger, we would be able to switch to a battery pack that attached to the behind-the-ear processor and would do away with the pocket and extra cable. Lisa started by attaching Alex's processor to her computer directly with an external cable. She then hooked the microphone behind his ear, though it hadn't yet been turned on, and placed the magnetic coil on his head.

"You try," she said, removing the coil and handing it to me.

Tentatively, I held the little piece of round flesh-colored plastic above his right ear. As soon as my hand drew near the internal magnet under his skin, the coil fell silently but energetically into position. How odd.

A snail-shaped outline appeared on Lisa's screen, a representation of Alex's cochlea and the electrodes that now snaked through it. Lisa clicked her mouse and a line of dots flashed on, one by one, along the curled inside of the snail. Each electrode was meant to line up with a physical spot in Alex's ear along the basilar membrane, so the electrodes were assigned different frequencies with high pitches at the base of the cochlea and low ones at the apex. The farther the electrodes are inserted, which depends on the patient's physiology and the surgeon's skill, the wider the range of frequencies that patient should hear.

"It's working," Lisa said. She had sent a signal to the internal components and they were functioning properly.

Next, Lisa pulled up a screen that showed each of the twenty-two electrodes that had been implanted inside Alex's cochlea in bar graph form. Two additional electrodes outside the cochlea would serve as grounds, creating a return path for the current. Lisa was going to set limits to the stimulation each electrode would receive—in effect, she was deciding how wide his window of sound should be. The process was known as MAPping, for Measurable Auditory Percept, and this was to be the first of many such sessions. At the bottom was the threshold, or T-level, the least amount of electrical stimulation that Alex could hear, and at the top was an upper limit known as the C-level, for "comfort" or "clinical," which would equal the loudest sound Alex could tolerate.

I stared at the computer screen, where each electrode showed up as a long, thin, gray rectangle. The sweet spot, the window of sound between the threshold and comfort levels, was depicted in canary yellow. Lisa could expand or compress the signal by clicking on the T or C level. Right there, in those narrow strips, was all the auditory experience of the world Alex was going to get in this ear.

We all looked at him. His eyes had widened.

"Can you hear that?" Lisa asked.

He nodded tentatively. Since Bill House's film of the young deaf woman Karen receiving her cochlear implant in the 1970s, there have been lots of videos of implant activations. YouTube has more than 37,000. Of the many I've seen, a few stand out. One ten-year-old girl burst into happy tears and clung to her grandmother, who, through her own tears, kept saying, "We did it, baby." A baby boy, lying in his mother's arms and sucking a pacifier, suddenly stared at his mother's face. His mouth dropped open in amazement, and the pacifier fell out.

Alex stayed true to his quiet, watchful nature. He stared at Lisa and then at me; his expression carried hints of surprise and uncertainty, but nothing more. The only drama was in my gut, tense with anxiety and hope. "It's working," I repeated to myself. Anticlimactic was okay, as long as it was working.

Because he was hooked up to the computer, I couldn't hear what Alex was hearing—he was having a private electronic conversation of clicks and tones. But he was responding. Sabrina had stacked toys on the table and was holding a basket. Just as in the sound booth, Alex was encouraged to throw a toy into the basket each time he heard anything. Kids, especially young ones, don't always report the softest sounds, so there's some art to assessing their responses. Once Lisa had set the levels to her satisfaction, she activated the microphone. Spinning her desk chair to face Alex, she ran through some spoken language testing, beginning with a series of consonants known as the Ling sounds. Developed by speech pathologist Daniel Ling, this is a test using six phonemes that encompass most of the necessary range of frequencies for understanding speech. Therapists, teachers, and parents use them as a quick gauge of how well a child is hearing.

"Mmmm." Lisa began with the low-frequency "m" sound, without which it's difficult to speak with normal rhythm or without vowel errors.

"Mmmm," Alex repeated. He had done this test many times at Clarke by this point.

"Oo," she said.

"Oo." He repeated the low-frequency vowel sound.

Then Lisa moved through progressively higher-frequency sounds. "Ee" has low- and high-frequency information. "Ah" is right in the middle of the speech spectrum. "Sh" is moderately high.

"Sss," she hissed finally. The very high-frequency "s" had been particularly hard for Alex to hear.

"Sss," he said quietly.

"It's working," she said again. "See how he responds over the next few days, and if he seems uncomfortable, let me know."

I put the fancy white Cochlear box, as big as a briefcase and complete with a handle, in the bottom of Alex's stroller. It contained a multitude of technological marvels: spare cables of various sizes, microphone covers, monitor earphones for me to check the equipment myself, cables for connecting to electronics, boxes of high-powered batteries. I thought back to my moment of ineptitude standing on the sidewalk in Brooklyn with Alex's brand-new, tangled hearing aids. That had been nine months earlier, and it was amateur hour compared to this.

"It will take a while," was the last thing Lisa told me as Alex and I left her office.

"It will take a while," Dr. Parisier had said when we saw us after the surgery. "His brain has to learn to make sense of what it's hearing."

Parisier had told us to use only the cochlear implant without the hearing aid in the other ear for a few weeks while Alex adjusted. But before too long, he wanted us to put the hearing aid back in the left ear.

"The more sound the better," said Parisier.

"Won't the two sides sound different?" I asked.

"They will," he said. "But he is two years old. I think his brain will be able to adjust. He will make sense of what we give him, and it will be normal for him."

When I reported Dr. Parisier's instructions about the hearing aid to Alex's teachers at Clarke, I realized just what unfamiliar territory we were in.

"He said that?" said one, a note of incredulity in her tone. "No one does that."

"They don't?"

She was not quite right, it turned out, but when I investigated, I could understand why she said it. Very few people were using both a hearing aid and a cochlear implant at that time. One reason was that in the early years, you had to be profoundly deaf in both ears to qualify for a cochlear implant. Most people who were eligible didn't have enough hearing in the other ear to get any benefit from a hearing aid. By 2006, when Alex began using his implant, that line of eligibility was shifting slightly. In the next few years, "bimodal" hearing—electrical in one ear, acoustic in the other—would become increasingly common and a major area of research. But I suspect Dr. Parisier's instructions for Alex were based more on instinct than science.

Sitting in the office at Clarke that morning, I realized that even though children had been receiving cochlear implants for more than fifteen years by then, Alex was still on the cutting edge of science. He did have more hearing in his left ear than was usual, at least for the moment. As long as that was true, it made sense to try to get the most out of that ear. But what would the world sound like to him?

We tried to damp down our expectations, but we didn't have to wait long to see results. On the day after Lisa turned on the implant, I was playing with the boys before bedtime. We had lined up their stuffed animals and were pretending to have a party.

"Would you like some cake?" I asked a big brown bear.

"Cake!" said Alex, clear as a bell. The "k" sounds in cake contain a lot of high frequencies. He had never actually spoken them before. Even Matty, at four, knew something important had just happened.

"He said 'cake'!" he exclaimed.

The following day, as I drove Alex up FDR Drive along the east side of Manhattan to Clarke, I let him watch *Blue's Clues* on our minivan's video system. "Mailbox!" I heard him call out in response to the prompt from the show's host. I spun around to look at him and narrowly avoided sideswiping a yellow cab.

"What did you say?!" He pointed at the screen and repeated "mailbox," though it sounded more like "may-bos." Of course, the show was designed to prompt him to say "mailbox." Jake and Matthew had done that. But Alex never had, despite watching that same *Blue's Clues* video many, many times.

15

A PERFECT STORM

At first, doctors thought eighteen-month-old Caitlin Parton had the flu. When they finally realized the toddler with her father's brown hair and her mother's blue eyes had meningitis, she was rushed to the hospital and stayed there for days. Doctors saved her life, but not her hearing. When she went home to her family's Manhattan apartment, Caitlin was profoundly deaf. Her parents, Steve Parton, an artist, and Melody James, an actress and director, were in shock. "People speak of the grief they feel on learning of their baby's hearing loss," James said years later. "For me, it felt like steep walls were suddenly in the path of our child's possibilities."

It was 1987. Fewer than three thousand people in the world had cochlear implants, nearly all of them adults who had lost their hearing after learning language. Those who'd heard before were thought to be more likely to succeed with the new device. "We didn't know how long their memories for sound would stay, but one assumed that they could tell us better what it was like," says Graeme Clark. For Clark and many other researchers, however, adults were just the beginning. "Deep down, I hoped it would help children," Clark says. "To give them an opportunity to communicate in the world of sound was really my life's work. But I couldn't do it until I'd done it in adults."

Whether a young brain that had never heard could make sense of what it received through an implant was a critical question. And it was one for which there was not yet a good answer. The explosion of studies in neuroscience and brain imaging lay a few years ahead. Helen Neville had just begun her studies of the brains of deaf and blind subjects in 1983. "Helping children was a completely different ball game," says Clark. "On the one hand, the theory was that [with] children, because their brains are supposedly plastic, you could give them anything and they might understand it. On the other hand, the contrary argument was: If they've never been exposed to sound, then these artificial electrical signals aren't going to be as good as the real thing. The dilemma was: Which hypothesis was correct?"

The stakes were different for children as well. "For kids, of course, what really counts is their language development," Richard Dowell told me. In addition to working with Rod Saunders and George Watson, Dowell had to figure out how to test the implant's viability in children. "In kids, you're trying to give them good enough hearing to actually then use that to assist their language development as close to normal as possible. So the emphasis changes very, very much when you're talking about kids."

Looming over the scientific question was the ethical question of whether it was right to subject young children to what amounted to experimental surgery. Clark's team tested issues of biological safety aggressively. But an implant for a child might be in place for decades, for a lifetime. Were the unproven possibilities worth the unknown risks—or even the known risks that accompany any surgery? Many clinicians were skeptical, sometimes angrily so. In 1984, one prominent otolaryngologist was quoted in *Medical World News* as saying, "There is no moral justification for an invasive electrode for children." He told the journal he found the cochlear implant a costly and "cruel incentive," designed to appeal to conscientious parents who may seek any means that will enable their children to hear. "It's a toboggan ride for those parents, and at the end of the ride is only a deep depression and you may hurt the kid."

For Bill House, it was meeting families of children who couldn't hear that had catalyzed his interest in deafness. Just as he hadn't let scientific and clinical skepticism stop him from operating on adults, he also pushed forward with children who'd lost their hearing to meningitis. As early as 1981, he had put his single-channel implant into the youngest person to be implanted to that point, three-year-old Tracy Husted. In 1984, two-year-old Matt Fiedor became the seventy-third child to get one of House's implants. His mother, Paulette, told me, "I was convinced that he had nothing, and if he could get any benefit from this device, something would be better than nothing."

There were no guarantees. Results to that point were extremely limited. Only about one in twenty recipients of any cochlear implant could carry on a conversation without speechreading. Consensus was growing, however, that the Bilger report had been correct and a multichannel implant was the most promising way forward. That view would be made official in a statement released following a 1988 conference convened by the National Institutes of Health.

After his implant was approved for adults in 1985, Clark began to work cautiously backward. That same year, he operated first on fifteen-year-old Peter Searle, then a ten-year-old named Scott Smith and, in 1986, five-year-old Bryn Davies. All three had lost their hearing as young children. The older the child, the longer he had been deaf. "The fifteen-year-old you could show some detection of sound but very limited benefit," remembers Dowell, who worked with all three boys, using some twenty-five speech production and perception tests—far more than are used today. The tests couldn't be very hard, since Dowell needed measures the kids might actually be able to achieve. He asked the boys to distinguish between one- and two-syllable words. He gave them closed-set and open-set words and sentences to repeat. "The ten-year-old was maybe a bit better." Bryn Davies, the youngest, had only been without hearing for two years when he got his implant at the age of five. Meningitis had ossified some of his ear canal and made it difficult to insert the electrodes very far. Nonetheless, his results were better than those of the two older boys. "He showed some promise," says Dowell, "enough to make you think, Aha! Now we're getting to somewhere." At that point, clinical trials, the round of research in which larger groups of children would get the implant and be closely watched and evaluated for several years, began.

In search of help, Caitlin Parton's parents had found their way to a New York City organization called the League for the Hard of Hearing, which offered information, speech therapy, and support groups. (Today, it is the Center for Hearing and Communication.) Because Caitlin had begun life with hearing, the Partons wanted to try an oral approach. Caitlin got hearing aids and began speech therapy, but the aids didn't help much. The League for the Hard of Hearing had been enlisted by Cochlear, the Australian company developing Clark's implant, and one of his collaborators, Dr. Noel Cohen of New York University Medical Center, to help find candidates for the clinical trials with children. As a preliminary step, the FDA had specified that the first clinical trial should include only children who had been born hearing and then become deaf, rather than those born deaf.

Caitlin was perfect.

"They said this device would give Catie a greater awareness of environmental sounds. That's all they promised us," said Steve Parton in an interview some years later, "that she might be able to hear cars honking and dogs barking. As parents living in New York City, that sounded pretty good."

It was no small thing, though, to have a child in an experimental trial. "There were no other families to talk with, no children to observe, no research studies to pore over and compare, no Internet, no listservs, no Twitter, no one to look at and speak to and share experiences with. There was no track record for the children," said Melody James. "It was like skating out on thin ice."

At the age of two and a half, Caitlin Parton became one of the youngest people in the world—and one of the first children that young in the United States—to receive a multichannel cochlear implant.

The winds of technological change were blowing. Facing into the gathering breeze, the Deaf community was determinedly flexing muscles it hadn't known existed, and the national media was paying attention. The spring of 1994 brought a new round of protests, this time at New York's Lexington School for the Deaf, a historic oral deaf school. When a hearing man named R. Max Gould was named chief executive officer of the Lexington Center, the institution that includes the school, Deaf leaders felt they had been left out of the search process. Wearing Deaf Pride T-shirts and carrying placards with messages like BOARD WHO CAN HEAR DON'T LISTEN, students and faculty organized days of protests at the school and at the offices of local politicians. After a week of pressure, Gould resigned and a deaf board president was installed to oversee the new search.

The Lexington protests were covered in a long feature in *The New York Times Magazine* by Andrew Solomon called "Defiantly Deaf." (It was this story that led to his 2012 book, *Far from the Tree: Parents, Children, and the Search for Identity*.) The year before, *The Atlantic Monthly* also ran an in-depth and much-talked-about article called "Deafness as Culture," exploring the central idea behind the movement: that deafness was not a disability. "The deaf community has begun to speak for itself," wrote author Edward Dolnick. "To the surprise and bewilderment of outsiders, its message is utterly contrary to the wisdom of centuries: Deaf people, far from groaning under a heavy yoke, are not handicapped at all." More than that, they were celebrating. At Lexington's commencement ceremony a few weeks after the successful protests, speaker Greg Hlibok, who had been one of the student leaders of the Deaf President Now movement, declared: "From the time God made earth until today, this is probably the best time to be Deaf."

There was a paradox here. Along with the spread of computers and the advent of e-mail, which radically improved communication for deaf people, an important reason it was good to be deaf at that moment in the United States was the passage of the Americans with Disabilities Act (ADA) in 1990. The ADA defines a disability as an impairment that "substantially limits a major life activity" and outlaws discrimination on the basis of disability in employment, education, transportation, telecommunications, and public accommodation (restaurants, theaters, sports stadiums, hotels, and the like). The telecommunications provision directly concerns people "with hearing and speech disabilities" and requires telephone companies to provide TTY, a keyboard system connected to telephones, and relay services that make use of a third party to allow deaf people to communicate by phone with hearing people. As a result, to take just one example, when a deaf person checks into a hotel today, he can expect the television to have captions, the phone to include TTY service, and flashing lights for the fire alarm. The ADA requires ASL interpreters in schools and at public meetings. Most provisions of the law were welcomed because they truly made it easier for deaf people to operate independently. As one commentator put it, the new law "leveled the playing field."

How could the Deaf reject "disability" as a concept that applied to them but accept the benefits of the Americans with Disabilities Act? It was an inconsistency that some found unsustainable. In a 1998 article for the nonpartisan bioethics research institute the Hastings Center, Bonnie Poitras Tucker, a disability law expert who is deaf herself, endorsed the provisions of the ADA as a way of allowing those with disabilities to take their rightful place in society. But with deaf people's newfound rights, argued Tucker, "come responsibilities." Citing the extensive cost of deafness—an estimated "$2.5 billion per year in lost workforce productivity; $121.8 billion in the cost of education; and more than $2 billion annually for the cost of equal access, Social Security Disability Income, Medicare, and other entitlements of the disabled"—Tucker made the argument (extreme to some) that "when most deafness becomes correctable . . . an individual who chooses not to correct his or her deafness (or the deafness of his or her child) will lack the moral right to demand that others pay for costly accommodations." (She added that cochlear implants were not likely to ever eliminate deafness altogether but claimed they might significantly reduce its "ramifications.")

Even in the Deaf culture camp, where views like Tucker's were anathema, some wrestled with the problem of how to reconcile the need for accommodations with their proud view of their experience. "Part of the odyssey I've made," a deaf adult named Cheryl Heppner told *The Atlantic Monthly*'s Dolnick, "is in realizing that deafness is a disability, but it's a disability that is unique." Others argued that since the law changed the environment, it provided access on deaf terms.

A cochlear implant, on the other hand, alters the person. And that, for many, was a problem. The Food and Drug Administration's 1990 decision to approve cochlear implants for children as young as two galvanized Deaf culture advocates. They saw the prostheses as just another in a long line of medical "fixes" for deafness. None of the previous ideas had worked, and it wasn't hard to find doctors and scientists who maintained that this wouldn't work either—at least not well. Beyond the complaint that the potential benefits of implants were dubious and unproven, Deaf culture advocates objected to the very premise that deaf people needed to be fixed at all. "I was upset," Ted Supalla told me. "I never saw myself as deficient ever. The medical community was not able to see that we could possibly see ourselves as perfectly fine and normal just living our lives. To go so far as to put something technical in our brains, at the beginning, was a serious affront." Waving his hand out the window at the buildings of George-town University Medical Center, where he is now employed, he gives a small laugh. "It's odd that I find myself working in a medical com-munity. . . . It's a real indication that times are different now."

The Deaf view was that late-deafened adults were old enough to understand their choice, had not grown up in Deaf culture, and already had spoken language. Young children who had been born deaf were different. The assumption was that cochlear implants would remove children from the Deaf world, thereby threatening the survival of that world. That led to complaints about "genocide" and the eradication of a minority group. Furthermore, implants would not necessarily deliver deaf children to the hearing world. Instead, the argument went, the children risked being adrift between the two, neither Deaf nor hearing. The Deaf community felt ignored by the medical and scientific supporters of cochlear implants; many believed deaf children should have the opportunity to make the choice for themselves once they were old enough; still others felt the implant should be outlawed entirely. "It felt like history repeating itself with a new vocabulary and new types of coercion," Carol Padden told me. Tellingly, the ASL sign developed for COCHLEAR IMPLANT was two fin-gers stabbed into the neck, vampire-style.

In the "For Hearing People Only" column of the magazine *Deaf Life*, the editors wrote:

> An implant is the ultimate invasion of the ear, the ulti-mate denial of deafness, the ultimate refusal to let deaf chil-dren be Deaf. Those who make the decision to implant

children choose to risk the children's health so that they can hear more sounds and develop clearer speech. Children attending oral schools and mainstream programs are the most likely to be implanted. Their parents, the ones who choose to have their children implanted, are in effect saying, "I don't respect the Deaf community, and I certainly don't want my child to be part of it. I want him/her to be part of the hearing world, not the Deaf world."

Deaf adults maintained that having lived without hearing themselves, they had something to offer deaf children that hearing parents could not. In the Deaf world, they said, deaf children could find a kinship that was at least as valuable as that of their biological families, if not more so. "Deaf people feel ownership of deaf children," Heppner said. "I admit it; I feel it, too. I really struggle in not wanting to interfere with a parent's right to parent and at the same time dealing with my own feelings and knowing that they have to accept that the child can never be one hundred percent theirs."

The message that hearing parents didn't know what was best for their own child did not resonate with the Partons. Neither did warnings of false hope and deep depression. Caitlin's mother and father wanted the rest of the world to hear about their experience. In November 1992, "Caitlin's Story" aired on *60 Minutes*.

Five years after her surgery, Caitlin was thriving. Her implant had indisputably delivered more than the sounds of barking dogs and honking horns. She was able to sit on the floor with Ed Bradley, who reported the segment, and carry on a meaningful spoken conversation about her cochlear implant with a total stranger.

"How does it work?" asked Bradley.

"The sound goes inside a tiny microphone around here," Caitlin said, pointing to the hearing aid–like device sitting on her ear. "It goes down into the computer, which is in this box." With her finger, she followed the cable that traveled from behind her ear to a purple zip case strapped to her waist. It contained the Walkman-size battery pack and speech processor, the external electronics where the speech signal was translated into an electrical code. As with all of the first generation of implants, Caitlin's processor and battery pack were bulky and had to be worn on the body. Then Caitlin's finger followed the wire from her waist back up to her ear and along a second wire, which connected the piece on her ear to a round plastic coil held to her head with a magnet. "Then it goes back up again here and then it goes into my brain."

"What happens if you take it off?" asked Bradley.

"Well, then I just hear nothing, just nothing. It's like this." Caitlin pursed her lips shut and made a face. "Nothing comes out."

The segment showed Caitlin in her mainstream classroom, singing a song with other students. She was shown in speech therapy and in hearing test booths. "You don't just put in an implant and hear as well as Caitlin does," Bradley noted. "Intensive speech therapy is as important as the device itself." For Caitlin, the combination of technology and hard work was paying off. Said Bradley: "She hears, sings, and talks like almost any of the other first-graders at [her] school in New York City." According to the program, her language skills—*60 Minutes* did not specify which ones—were two years ahead of her age, and her comprehension skills were three years ahead.

Not everyone was as wowed as Bradley.

"The impression the child gets with all of that focus on the ear, on the mouth, is that maybe the parents would love me more if I could just hear a little better, if I could just talk a little better instead of Mom accepting me just the way I am," said Roslyn Rosen, then the president of the National Association of the Deaf, when Bradley interviewed her. The year before, the NAD had released an official statement, written by psychologist Harlan Lane, "deploring" cochlear implants for kids and saying: "[It is] invasive surgery on defenseless children, when the long-term physical, emotional, and social effects on children from this irreversible procedure—which will alter the lives of these children—have not been scientifically established." At the time of the *60 Minutes* broadcast, the NAD was lobbying to have the FDA's approval of cochlear implants for children revoked.

Rosen also spoke quite lyrically about her experience of the world. "Many people have said to me, 'What is it like to be a deaf person? You don't hear the birds sing, the leaves rustling as the wind breezes through the trees, the crash of the ocean as it hits the shore. Aren't you missing those things?'" she said through an interpreter. "All of that is music to my eyes. Deafness is like a prism that plays up brilliantly those things that may be missing to other people."

After the show aired, Caitlin became a symbol of the wonders of technology—hers is still a well-known name in cochlear implant circles. Melody James called the show "the shot heard round the world." But *60 Minutes* also received angry letters from Deaf people accusing the program of participating in "child abuse" and "genocide."

Confrontations between the Deaf community and the medical community became a regular occurrence at international conferences. In France, a group called Sourds en Colère (Angry Deaf) tried to disrupt meetings in Lyon and Paris, and a young man who had received a cochlear implant but turned against the device was cheered as he smashed his processor to bits with a sledgehammer on the sidewalk. In Manchester, England, the Deaf Liberation Front laid empty coffins in the street in front of the building where the International Cochlear Implant Conference was being held and waved banners saying BETTER DEAF THAN DEAD. This was a reference to the slightly increased risk of meningitis caused by implants, particularly by a device used in Europe that included two parallel electrodes with a space between them that proved particularly prone to infection. After six implanted children died of meningitis in Europe, that device was recalled and the cases of meningitis were much reduced, but the Deaf community was angrier than ever. In Melbourne, protesters filled Collins Street in front of the hotel where Graeme Clark had organized a conference. Their message was: SAY NO TO COCHLEAR IMPLANTS. Clark remembers, "I was written up as nearly evil."

Tension between the emotion of parents and the emotion of a newly politicized Deaf community bubbled up in unusual places. Debi and David Leekoff's son, Mark, received a cochlear implant at the age of four as part of the clinical trials. Several years later, Debi took Mark to see a taping of the game show *Jeopardy!* near their northern Virginia home. Outside, Mark and his mother got in the line for preferential seating, which included anyone who was deaf or hard of hearing. Also in the line was a group of students from Gallaudet, with one person who could speak. When they spotted Mark's cochlear implant, they began to yell for the manager.

"You're not deaf," the group's spokesperson declared. "You should go home."

"He needs to be taken out of this line," the students told the manager. "He's not deaf!"

"It was horrible, horrible," remembers Debi. "Mark cried. He wanted to go home. Talk about bullying!"

The researchers and doctors around the world who were working on cochlear implants believed in their cause, too. They had mixed reactions when they ran up against demonstrators at conferences. Rob Shepherd, who was responsible for safety testing on implants for children in Melbourne, said of the wish to maintain a culture of sign language, "I didn't agree with it, but I could see where it was coming from." Others were more vehemently angry in response. "It's child abuse *not* to implant a child," one audiologist told me. Dr. Parisier,

Alex's surgeon whose family fled France in the face of the Holocaust, was deeply offended. "I take the word 'genocide' very seriously," he says.

These doctors and researchers also spent a lot of time with another group of deaf people—those who did want to hear. Bonnie Poitras Tucker, who wrote about the ADA, certainly did not subscribe to what she called the "deaf is dandy" camp. "Most of us would *love* [emphasis original] to be able to pick up the telephone and make a personal or business call when and how we feel like it without having to scramble to find an interpreter and without having to make the call with a third person privy to every word," Tucker wrote. "We'd like to be able to go to a movie or a play regardless of whether captioning or interpreters are available. We'd like to be able to participate in group conversations, to hear the conversation at the dinner table. We'd like to be able to hear music; to hear our children and grandchildren laugh and cry; to listen to the radio when we are driving; to have a car phone; to be able to use the drive-up window at McDonald's; to hear the announcements at the airport; to be able to talk to the person in front of or behind us on a hiking trail; to be able to go to a professional meeting on the spur of the moment; to be able to get any job we want without having to consider how our deafness will interfere with the job duties. We'd particularly like to hear our own voices and to be able to control the tone and pitch and loudness of our voices."

Like Tucker, those who had succeeded with oral education—and there were plenty—didn't see themselves reflected in all the talk of Deaf culture. Some tried to mount a countercampaign. In the early 1990s, word got out that the Smithsonian Institution in Washington, DC, was planning an exhibit called "Silent America," which was intended to help raise awareness of Deaf culture and American Sign Language. The oral deaf community resented that their experience was left out of the exhibit. Angry letters flew between the two camps and to the organizers. Eventually, the Smithsonian gave up on the idea. That was a result, noted one group of deaf leaders, that meant "regardless of what side anyone was on, we all lost that battle."

In May 2006, four months after Alex's cochlear implant was activated, I sat down at the kitchen table one morning with *The New York Times* and was surprised to see that Gallaudet University was in the news again. In 1988, when Deaf President Now occurred, I had been a senior in college studying American social history. I was aware of the protests as an interesting development, but that was all. This time, I read the story with rapt attention. Now it was personal. Would these people play a role in my life? It was a little like considering the parents of a new boyfriend as potential in-laws.

Once again, I read, students were protesting. Once again, they were dissatisfied with the way in which a new university president had been chosen. I. King Jordan, the deaf man who had taken the helm in the wake of Deaf President Now, was stepping down after eighteen years. His protégé, Provost Jane Fernandes, had been named by the board of trustees to replace him the following January.

She was an unpopular choice. Soon after the board announced Fernandes's appointment, the faculty issued a vote of no confidence in her. They and the students said the search process was flawed and opaque. The demonstrations carried not the righteous anger of 1988, born of newfound pride and political coming-of-age, but an anger that was ugly and divisive. Bomb threats kept the gates locked on commencement day. The chair of the board, citing "aggressive threats," stepped down a week after the decision to appoint Fernandes. Students were quoted in the paper making very personal complaints about Fernandes: She was aloof; she signed poorly; she married a hearing man.

Fernandes, who had grown up deaf but hadn't learned to sign until she was twenty-three, responded by saying that the protesters were locked in a narrow view of deafness and that in their eyes, she was "not deaf enough." A flyer pasted around campus, however, said: "It's not that she's not deaf enough. She's not enough of a leader." Looking at the small tents protesters had pitched on the lawn, Fernandes told a *New York Times* reporter that her vision for the university was as "one big tent for all." Recognizing that the deaf community was facing change from cochlear implants and mainstream education, she argued that Gallaudet had to embrace "all kinds of deaf people." ASL would always be central to Gallaudet, but the university had to welcome people who hadn't grown up signing as well. "We're in a fight for the survival of Gallaudet University," she said.

The conflict carried on for months, quieting in the summer and then exploding again in the fall, and I continued to follow it closely. In October, more than one hundred students were arrested for taking over a campus building, and a few began hunger strikes. As in 1988, the protests finally shut down the university for several days. The clash, noted one reporter, "is illuminating differences over the future of deaf culture writ large, and focusing attention on a politically charged debate about what it means to be deaf in the 21st century."

That was certainly a question I was asking myself.

At the end of October, the protesters won again—at least on the surface—when the board revoked Fernandes's appointment. Students jubilantly celebrated on the university's football field, home of the team credited with inventing the huddle so that their opponents couldn't see their signs.

I was left unsure as to what had really happened. The victory seemed much less clear than in 1988. I had no way to judge Fernandes other than through the news reports, but the demonstrations themselves seemed as problematic as the process they were protesting. Were students going to shut down the university every time they didn't like a board decision? Pragmatically speaking, Fernandes had a point. The pace of technological innovation was unrelenting. It was forcing change on the deaf community. Would a child like mine ever want to go to Gallaudet or another school for the deaf? Would he have to reject his cochlear implant to do it? The devices were so unpopular on campus that students (and professors) were said to feel they couldn't wear them there.

At the culmination of the protests, I read a comment from Lawrence Fleischer, then-chairman of the Deaf Studies Department at California State University, Northridge, another important center of deaf higher education. "More parents are choosing cochlear implants for their children," he said. "We call it the false hope. We call it the magical consciousness, meaning that their consciousness is way below average, but they're pretending to have consciousness they don't really have." I wanted to like Deaf culture, but his comment left me cold. Granted it had been less than a year since Alex had begun using his implant, but "false hope" was so contrary to our experience that it was hard to take it seriously. Alex was making steady progress. I was willing to accept that success was variable, but I knew Alex was not alone. To dismiss anyone who did well seemed like willful ignorance.

At least one truth behind the protests emerged later. Gallaudet was a university in trouble and not just because of the cochlear implant. King Jordan was leaving a mess. Though he was once a hero, the faculty voted no confidence in him as well that fall. In November 2006, one month after Fernandes's appointment was revoked, the Middle States Commission on Higher Education, the body responsible for accrediting degree-granting colleges and universities, expressed "serious concern" about the state of affairs at Gallaudet and postponed a decision on reaccreditation. For the next two years, the school was on probation. Concerns included not just the recent presidential search process but also the need to nurture a climate of respect among students, faculty, and administrators, the need for a strategic plan, evidence of academic rigor, and more rigorous reporting to the commission. The numbers spoke for themselves. Less than 30 percent were graduating in six years. The following year, 2007, only slightly more than half even made it to sophomore year. Most students needed remedial help in English and math. As a result of the protests, the probation, and a subsequent effort to tighten admission standards, undergraduate enrollment dropped from 1,600 in the mid-1990s to 1,080 in 2007. Fernandes had been right about one thing: Gallaudet, and perhaps by extension Deaf culture, really was fighting for its survival.

Discouraged, I wondered anew about the reasons for such pervasive academic struggles. For Alex, for the time being, I resolved to stay focused on sound.

16

A Cascade of Responses

Dressed in a cotton smock, stripped of my earrings, watch, belt, and all other sources of metal, I am lying on a blue-sheeted bed with my head cradled in what looks alarmingly like a heavy-duty toilet seat. Above me, through an opening in the thick, white, fiberglass C that surrounds my head, I can see a computer screen booting up. It's actually the projected image of a monitor mounted horizontally so it can be seen from the bed. A real computer is electrical and can't be in here with me.

"Are you okay in there?" a man calls to me through a speaker.

"Just great," I answer.

A pause. "We are watching." A small laugh.

Deep inside the second floor of New York University's Department of Psychology, I have become a research subject. I came to see David Poeppel, whose interests as a researcher align almost perfectly with mine as Alex's mother, since Poeppel studies both sound and language. Lanky and boyish except for the sprinkling of gray in his hair, and usually dressed in jeans and black-framed glasses, Poeppel is two parts thoughtful MIT-educated linguist and neuroscientist and one part enthusiastic teenager who happens to have some really cool gadgets. He loves to wrestle with big ideas and their backstories. Happily for me, he loves to talk about ideas, too, in the kind of wide-ranging conversation that puts "blob" and "interdigitate" in the same sentence. Within twenty minutes of meeting me, Poeppel decides that the best way for me to understand what happens to sound when it reaches the brain is to see it for myself. "We can record your own hearing," he says. And then he puts me inside the MEG machine.

MEG, or magnetoencephalography, "one of the new boutique brain-imaging approaches" as Poeppel puts it, has him and his colleagues as thrilled as technophiles with their first iPads. At a recent conference, Poeppel exhorted the audience to "run, don't walk, to the nearest MEG scanner." The technology provides something approaching the spatial resolution of MRI—which tells us, says Poeppel, "what is here versus here versus here, but it's not a theory or mechanism of how hearing and speech and language work, because they happen incredibly fast"—and it gives the temporal resolution (timing) of EEG, which "gives you millisecond by millisecond of what's happening in your head, but you can't figure out where it's coming from." With MEG, neuroscientists don't have to choose between timing and location; they can have both.

Wherever there is an electrical current, there is also a magnetic field curling around it. You can use the "right-hand rule" to determine its direction: Stick out your right thumb as if to hitchhike and curl your fingers toward your palm, and you have created a model of the electrical current—which moves in the direction your thumb is pointing—and the magnetic field, which follows your fingers to spin around the current. Magnetoencephalography measures the direction and intensity of the magnetic field surrounding the electrical one, thereby locating the current. In your brain, that is a feat roughly equivalent to recording a pin dropping while an orchestra is playing.

Like an MRI machine, an MEG scanner is kept in a shielded room, but for opposite reasons. MRI artificially generates a large magnetic field that is potentially dangerous. Unleashed, it would erase credit cards and stop pacemakers. I once saw a technician demonstrate its power by holding a sheet of metal at the door of the MRI room. The square foot of metal flipped up instantly and would have flown across the room like the *Millennium Falcon* in the grip of the Death Star's tractor beam had the technician not held on with two strong hands. So the shielded walls are there to keep the magnetic field in. (Alex's implant means he cannot undergo MRI.)

At NYU, the MEG machine is inside a room whose walls, floor, and ceiling are about ten inches thick, with three layers of copper and a layer of a special alloy called mu-metal—all there because "the signal that we're measuring in your head is so small, so unbelievably small, that we need to be keeping all the electricity—the elevator, the subway, the traffic—in our world out," says Poeppel. It's not just subways and elevators that obscure the signal; even the electricity in your body interferes. "Your heart generates a huge signal like the music at the beginning of *Law & Order*: Dunh, dunh. That's a gigantic electrical spike. Of course, since it's an electrical spike, it's a magnetic spike. We don't want that signal, since we're studying the head. Even the heart is really big compared to what we're trying to get from the head."

He is warming to his subject.

"So here's what's cool cocktail party conversation . . . well, depends on the party," he admits. "The range that we're recording this is femtotesla." A tesla (T), named after engineer and physicist Nikola Tesla, is the unit of measurement that describes the strength of a magnetic field. The superconducting magnet built around the CMS detector at CERN on the French-Swiss border to search for the Higgs boson particle is 4 T. Most MRIs range from 1.5 to 3 T, and a refrigerator magnet is about five millitesla. A femtotesla is a quadrillionth of a tesla. "Ten to the minus fifteen," explains Poeppel. "Formally speaking, wicked small."

Inside the toilet bowl–shaped head cradle of this particular MEG machine, there are 160 detectors called SQUIDs, which stands for superconducting quantum interference devices. (There are also seated versions, with a helmet like an upside-down toilet bowl, that contain even more SQUIDs.) Each SQUID is the size of my pinkie and consists of a coil of a special wire suspended in liquid helium.

Lying on the flatbed of the MEG, I don't have to do anything at all. "If your brain is in there and it can hear, we'll see it," says Poeppel. He and his MEG technician, Jeff Walker, are sitting outside the shielded room at a bank of computers. Walker has inserted earbuds in my ear canals.

"Do I have to lie still?"

"If you squiggle your head around, the data suck."

"I'll lie still."

"Ready?"

"Ready."

I hear beeps. The same tone repeats. Beep, beep, beep. This goes on for a while. Then a second, higher-pitched tone begins. Beep, beep, beep.

Poeppel describes this two-tone test as one of the "most boring tests there is."

It's pretty boring. I am very purposefully doing nothing, not even concentrating on the beeps.

But when we're done and I join them at the computers, I can see from the results that my brain was plenty busy.

"There is your brain on sound, actually," says Poeppel. "We gave you five minutes' worth of tones—high tones and low tones. I know you were fascinated."

For each tone, there are two graphs on the screen showing my auditory response. The first is a waveform map, with a tangle of sinewy black lines, one for each detector, undulating from left to right. The second graph shows the movement of the magnetic field. It's called an isofield contour map and features a stick-figure outline of a head, with a triangular nose tacked on top and half-circle ears on the sides to show we're looking down from above. Here, small blue diamonds dotting the cartoon head represent the detectors. With some quick adjustments, Walker cleans up the data and separates it so we can begin to look for patterns.

"This is just the high tone," says Walker, once he has created a neater graph with slightly fewer lines. The tone was at 1,000 Hz and I heard it one hundred times, but a few of those times the signal "had some sort of wonky weirdness, like maybe a channel went crazy or you coughed." Clearing out the wonky weirdness left us with eighty-four lines. The undulations are more individual now, their separate characteristics clearer in places, but still a mass of black in the middle. The x-axis of the graph shows milliseconds, and the y-axis shows the wicked small femtotesla. Beginning at the start of the beep, which scientists call the onset, the lines roll along gently until about eighty milliseconds, when they suddenly begin to loop above and below the center line, as if bursting into a mass game of double Dutch, with the top of the jump rope arcing around 120 milliseconds. Then the lines slacken, stop their game, and amble along again with just enough wiggle to suggest that someone is still lightly flicking a wrist at the end of the rope.

I remember seeing a similar peak in the electrical waveforms that Anu Sharma showed me at her laboratory in Colorado where she studies the first valley (the P1) and, in older children, the first peak (the N1) in EEG recordings.

"That is the N1, and you have one," Poeppel confirms. "So good news."

In MEG terms, he adds, the response is called the M100 or the N1m. Whatever the name, it represents the same thing: confirmation that the auditory cortex is processing sound. The sound doesn't have to be natural. It can be conveyed by hearing aids or cochlear implants.

The N1 takes some time to develop and is seen only in older children and adults. But the P1 shows up in very young children. That is precisely why it is useful to Sharma. As she had explained, this particular early response matures as a child ages, becoming faster and more efficient. By measuring how quickly the P1 appears, Sharma can tell whether a child is getting enough sound to develop spoken language long before that child says her first word. The technique can save precious time if hearing aids aren't going to be effective.

Researchers like Poeppel are trying to define what happens not in young children but in adults who are experienced listeners. They spend a lot of time on the N1 or its magnetic equivalent because, as Poeppel puts it, "it's big ticket"—easy to see and packed with information. By glancing at an N1 response, for instance, someone who's practiced at reading such data can tell whether the subject heard an "ah" or an "ee," or whether she was paying attention. "You can unpack the nature of hearing and perception from this very elementary response," says Poeppel.

The two graphs from my test are linked on Walker's computer screen. By moving the cursor along the lines of the waveform map, he can replay the movement of the magnetic field in my head on the contour map. As he does, concentric splotches of red and blue, depicted like a topographic map with the deepest colors at the center, change shape and intensity as they shift across the landscape of my brain. "This is the flat map of the thing," says Poeppel. "You have to imagine it curved around your head." The magnetic field in the brain is always present. Walker describes it as "a constant fluctuating ocean of activity." And it's large enough that it passes in and out of the brain as it moves. I imagine a ghostly halo revolving around and through my skull. On the map, blue represents the "sink," the magnetic flux coming into the head, and red is the "source," the magnetic activity coming out of the head. When a beep sounds, it's like throwing a rock in that ocean of activity. Things change. The colors get deeper; the splotches concentrate and then swap places. Blue moves from front to back and red from back to front. Although there's color all over, the activity seems most intense on the left, which is as it should be.

"These are the temporal lobe channels for a sound. If we don't see those, we're not so happy," says Poeppel. "You're generating, happily, a very stereotypical boring pattern."

"What does the movement around the head represent?" I ask. "What are we seeing?"

"Well, it changed direction. The current was shooting in one direction, presumably communicating information: 'I was a tone of the following type.' The next one is already shooting information a different direction."

"Okay, so the electrical current is zinging around my brain," I say. "But what does it mean from one shift to the next? What does it mean when it changes direction?"

"The contour map is like a mountain range. Is the magnetic field going up the mountain or down into the ocean? It tells you the underlying brain activity is changing its direction."

But on a deeper level, he says, we still don't know what that means.

"This is what we're working towards," says Poeppel. "If I knew the answer to what this means, I'd be an exceptionally famous person."

When I compare my MEG recordings with Alexander Graham Bell intoning vowels into tuning forks and Harvey Fletcher at Bell Labs having M.A. (male, low-pitched) and F.D. (female, high-pitched) speak the word "farmers" again and again, it's obvious our understanding of sound and hearing has come a tremendous distance. Why, I wonder, do so many of the researchers I've talked to emphasize how much we don't know? It's such a common refrain that while writing this chapter, I heard a radio interviewer ask esteemed neuroscientist Eric Kandel of Columbia University what mysteries remain about the brain. "Almost everything," answered Kandel.

"What's going on?" I ask Poeppel. "Surely we do know more."

"Look, I work on problems that were all defined in the nineteenth century. Most of what we know is footnotes to Helmholtz," he says, referring to the German physicist Hermann von Helmholtz, who inspired Bell.

Then he pauses.

"No, that's not quite right. We have made a lot of progress, but it's in part because the finer-grain you look, the more new things come up." As an example, he reminds me that more than one hundred years ago, Camillo Golgi and Santiago Ramón y Cajal, who shared a Nobel Prize for their foundational work in neuroscience, engaged in a great debate over whether the neuron was really the unit of operation in brain activity.

"What they were looking at were these absolutely gorgeous pictures of cells with these long protrusions, and they had the best microscopy that was available," Poeppel says. "They said that must be the unit we care about." Fifty years later, scientists discovered the synapse and that became the focus. Then another thirty years passed. "Now, suddenly, what turns out to be the unit of modification for synaptic plasticity are dendritic spines, which are a fraction of a fraction of a cell," says Poeppel. No one could see them before. "It's not that we don't know; we do know more. The problem is: Every time we go to higher resolutions, we're discovering in part new phenomena and in part how we have to reinterpret old phenomena."

The explosion in brain-imaging technology—the "supercool" MEGs and fMRIs and PETs—made this new era of granularity possible by allowing scientists to record the activity of tens of thousands of cells firing at the same time. Until the mid-1990s, most data on the brain came from injuries, deficits, lesions, and so on. Data from deficits, though, would never have given you what Poeppel and Walker produced in just five minutes of watching my brain on sound, or what they could get if we had carried on with speech sounds. "I can stick you in there right now and we can do 'ah's,' 'oo's,' and 'ee's' all day long until we map your space," says Poeppel. "We can map the timing, we can do the spatial analysis. We can do a 3-D reconstruction. Everyone's a little different, but there's a high degree of consistency. Any auditory stimulus will have this cascade of responses."

In other words, everything from a beep to a recitation of *Macbeth* sends waves of electrical pulses rippling through the brain along complex though predictable routes on a schedule tracked in milliseconds. Along the way, the response to a beep gets less complicated because a beep is, well, less complicated than *Macbeth's* guilty conscience. But any word from "apple" to "zipper" takes less than half a second to visit all the lower and higher processing centers of the brain, and in that fraction of a second neuroscientists have a pretty good idea of the several stops the word makes and the work done by the brain along the way. I now understood how the P1 and N1 got their names. Each point where the response is distinctive and concentrated has been labeled with an N or a P depending on whether the wave is usually negative-going or positive-going at that point (counterintuitively to me, convention dictates that negativities go up and positivities go down), and then the N or P is given a number to indicate either how many milliseconds were required to get there or how many major peaks have preceded it. Taken together, these responses have helped us understand far more about the steps in the intricate dance the brain performs as it converts sound to language.

In a hearing adult, in the first fifteen milliseconds, the signal, or sound, is in the brain stem. In this early and relatively small section of the brain, the sound will be relayed through the cochlear nuclei, the superior olive, the lateral lemniscus, and the inferior colliculus in the midbrain, subdividing and branching each time onto one of several possible parallel pathways, until it reaches the early auditory cortex. "What has been accomplished so far?" asks Poeppel. "All these nuclei have done an enormous amount of sophisticated analysis and computation." The superior olive, for instance, is the first place where cells are sensitive to information coming from both ears and begin to integrate and compare the two, the better to figure out where the sound originated. "Doing things like localizing . . . is not necessarily accomplished, but the critical agreements are already calculated for you at step two," says Poeppel. "It's pretty impressive." It was these early distinctive responses that Jessica O'Gara was looking for in Alex when she gave him an auditory brain stem response test, designed to see if the auditory path to the brain is intact.

If the route is clear, as it is in a typically hearing person, the sound reaches the primary auditory cortex in the temporal lobe somewhere around twenty milliseconds and spends another few milliseconds activating auditory areas such as Heschl's gyrus. The P1 is usually at about sixty milliseconds, and the N1, as Poeppel and Walker just recorded in me, peaks around one hundred milliseconds. It's still mainly an auditory response, as the brain continues to identify and analyze what it just heard, but the brain is also starting to take visual information into account. Already, at one hundred milliseconds, higher-order processing is under way. The more unpredictable the sound—a word you rarely hear, for instance—the bigger the amplitude of the N1 because the harder you're working to make sense of it. Conversely, the response is more muted for words we hear all the time or sounds we expect, such as our own speech.

Around two hundred milliseconds, where the P2 or P200 occurs, the brain starts to look things up. It is beginning to compare the arriving sound with what it already knows by digging into stored memories and consulting its mental dictionary. As befits this more complex task, the signals are now firmly in the higher regions of the brain, and the neural activity is far more widespread. This is also the point where information arriving from other systems truly converges. However a word is perceived—whether you hear it, read it, see it, or touch it— you will begin processing it fully at this same point.

The ability to recognize words, to acknowledge the meaning found in the dictionary, is known as lexical processing and happens between two hundred and four hundred milliseconds, which is very late in brain time. As was true earlier, the amplitude of the N400 reflects the amount of work the brain is being asked to do: It is larger for more infrequent or unfamiliar words and for words that vary from many others by only one letter, such as "hat," "hit," "hot," and "hut." By six hundred milliseconds, the brain is processing entire sentences, and grammar kicks in. The P600 is thought to reflect what neuroscientists call "repair and reanalysis," because it is elicited by grammatical errors and by "garden path sentences" (the kind that meander and dangle their modifiers: "The broker persuaded to sell the stock was tall") and—in a non-linguistic example—by musical chords played out of key.

But all of that is only half the story. The cascade of responses that begins with the ear and leads all the way to the ability to follow a poorly constructed sentence—to know that it's the broker who is tall— is known as bottom-up processing. It starts with the basic input to any sense—raw data—and ends with such higher-level skills as reasoning and judgment and critical thinking—in other words, our expectations and knowledge. Neuroscientists now believe that the process is also happening in reverse, that the cascade flows both ways, with information being prepared, treated, and converted in both directions simultaneously, from the bottom up and from the top down.

This idea amounts to a radical rethinking of the very nature of perception. "Historically, the way we intuitively think about all perception is that we're like a passive recording device with detectors that are specialized for certain things, like a retina for seeing, a cochlea for hearing, and so forth," says Poeppel. "We're kind of a camera or microphone that gets encoded somehow and then magically makes contact with the stuff in your head." At the same time, many of the big thinkers who pondered perception, beginning with Helmholtz (him again), knew that couldn't be quite right. If we reached for a glass or listened to a sentence, didn't it help to be able to anticipate what might come next? In the mid-to-late twentieth century, a handful of prominent researchers proposed models of perception that suggested instead that we engaged in "active sensing," seeking out what was possible as we went along. Most important among these was Alvin Liberman at Yale University, whose influential motor theory of speech perception fell into this category. He proposed that as we listen to speech, the brain essentially imagines producing the words itself. Liberman's elegant idea and other such ideas did not gain much traction until the past decade, when they suddenly became a hot topic of conversation in the study of cognition. What everyone is talking about today is the brain's power of prediction.

That power is not mystical but mathematical. It reflects the data-driven, statistical approach that informs contemporary cognitive science and defines the workings of the brain in two ways: representations and computations. Representations are the equivalent of a series of thumbnail images of the things and ideas we have experienced; everything in our mental hard drive, like the family photographs stored in your computer. How exactly they are stored remains an open question—probably not as pictures, though, because that would be too easy. Computations are what they sound like: the addition, subtraction, multiplication, and division we perform on the representations, as if the brain begins cropping, rotating, and eliminating red-eye. They are how we react to the world and, crucially, how we learn. "The statistical approach makes strong assumptions about what kinds of things a learner can take in, process, 'chunk' in the right way, and then use for counting or for deriving higher-order representations," says Poeppel.

On one level, prediction is just common sense, which may be one reason it didn't get much scientific respect for so long. If you see your doctor in the doctor's office, you recognize her quickly. If you see her in the grocery store dressed in jeans, you'll be slower to realize you know her. Predictable events are easy for the brain; unpredictable events require more effort. "Our expectations for what we're going to perceive seem to be a critical part of the process," says Greg Hickok, a neuroscientist who studies predictive coding among other things at the University of California, Irvine, and regularly collaborates with Poeppel. "It allows the system to make guesses as to what it might be seeing and to use computational shortcuts. Perception is very much a top-down process, a very active process of constructing a reality. A lot of that comes from prediction."

Predictive coding has real implications for Alex, Hickok points out. "Someone with a degraded input system has to rely a lot more on top-down information," he says. "If you analyze sensory input roughly, you test against more information as it's coming in. Let me look and see if it matches." Anyone who reads speech is using prediction, guessing at context from the roughly one-third of what is said that can be seen on the mouth and using any other visual cues he can find. Those who use hearing aids and implants still have to fill in gaps as well. No wonder so many deaf and hard-of-hearing children are exhausted at the end of the day.

But top-down processing can be simple, too. If a sound is uncomfortably loud, for instance, it is the cortex that registers that fact and sends a message all the way back to the cochlea to stiffen hair cells as a protective measure. The same is true of the retina, adjusting for the amount of light available. "It's not your eye doing that," says Poeppel. "It's your brain." Then he beats rhythmically on the desk with a pencil: tap, tap, tap, tap. "By beat three, you've anticipated the time. By beat four, we can show you neurophysiologically exactly how that prediction is encoded."

"Helmholtz couldn't do that," I point out.

"We have pretty good theories about each separate level," Poeppel agrees, citing the details of each brain response as an excellent example of fine-grained knowledge at work, but "how is it that you go from a very elementary stimulation at the periphery to understanding in your head 'cat'? We don't know. We're looking for the linking hypothesis."

I realize it's the same question that Blair Simmons asked at the cochlear implant workshop in 1967: How do we make sense of what we hear? And I have to acknowledge that, as Eric Kandel said, mysteries remain.

What kind of information does a sound have to carry in order to set this auditory chain in motion? That was another question Simmons asked. With Alex in mind, I wonder what happens if *Macbeth* is too quiet or garbled or otherwise distorted. I put this basic question to Andrew Oxenham, an auditory scientist at the University of Minnesota, who studies the auditory system in people both with and without hearing loss.

"What cochlear implants have shown us," he tells me, is just how well we can understand speech in "highly, highly degraded situations. Think about the normal ear with its ultrafine-frequency tuning and basilar membrane that [has] thousands of hair cells, and you think about replacing that with just six or even four electrodes—so we're going from hundreds, possibly thousands, of independent frequency channels down to four or six. You'd say: Wow, that's such a loss of information. How is anyone ever going to perceive anything with that? And yet people can understand speech. I guess that's shown us first of all how adaptable the brain is in terms of interpreting whatever information it can get hold of. And secondly, what a robust signal speech is that you can degrade it to that extent and people can still extract the meaning."

This makes sense, Oxenham adds, from an evolutionary point of view. "You want something that can survive even in very challenging acoustic environments. You want to be able to get your message across." Engineers often think about redundancy, he points out, and like to build things with a belt-and-suspenders approach, the better to ensure success. "What we have learned is that speech has an incredible amount of redundancy," says Oxenham. "You can distort it, you can damage it, you can take out parts of it, and yet a lot of the message survives."

Just how true that is has been shown in a series of experiments designed to test exactly how much distortion and damage speech can sustain and still be intelligible. Back in David Poeppel's office at NYU, he walks me through a fun house of auditory perception to show me what we've learned about the minimal amount of auditory information necessary for comprehension.

He begins with sine waves, a man-made narrow-band acoustic signal that lacks the richness and texture of speech—a beep, really. On the whiteboard in his office, Poeppel sketches out a spectrogram, a graph depicting the range of frequencies contained in sound over time, with squiggly lines representing bands of energy beginning at 100, 900, 1,100, and 2,200 Hz and then shifting.

"This is me saying 'ibex,'" he says.

"If you say so," I answer. I wouldn't have known the details, but by now I know those bands of energy are the defining formants—the same features that Graeme Clark and his team used to design their first speech processing program.

"Do you really need all that spectral information to extract something intelligible?" Poeppel asks.

The answer is no. Some of the spectral information, the combination of frequencies that makes up each sound, turned out to be icing on the acoustic cake. In the early 1980s, several researchers, notably Robert Remez at Barnard University and Philip Rubin of Haskins Laboratories, created spectrograms of spoken sentences, got rid of the fundamental frequency (the band at 100 Hz in "ibex"), extracted the formants (the remaining bands), and replaced them with sine waves. With sine waves standing in for formants stacked up at the appropriate frequencies, the sentence is intelligible. If you practice, that is. Poeppel plays me some samples from the Haskins Laboratories website. At first, I just hear high-pitched whistles like the sounds Jodie Foster interpreted as alien communication in *Contact*. But after Poeppel plays the natural sentence and then replays the sine-wave sentence, I hear it: "The steady drip is worse than the drenching rain." Do this for twenty minutes, Poeppel tells me, and you'll get really good at it.

At the House Research Institute in Los Angeles, an auditory neuroscientist named Bob Shannon and his colleagues took a different approach. They manipulated sentences not by minimizing the spectral information but by degrading it. When the ear groups similar frequencies into critical bands, or channels, along the basilar membrane, there are typically about thirty in a healthy ear. In an influential experiment, Shannon and his team took a set of sentences and reduced the number of channels used to convey them, as if thirty radio stations were trying to broadcast from just a few spots on the dial. Condensed into one big channel, the sentences were basically unintelligible. With two channels, there's minimal information, but still not enough to understand much. With three channels, the sentences coalesce into words. "The difference between two and three is huge," says Poeppel. "At four, you're almost at ceiling." In other words, when asked to repeat what you heard, you would get nearly every sentence even without the remaining twenty-six channels of information about sound you normally use.

What if you leave the spectral composition of sentences alone and change only the timing, what scientists call the temporal information, instead? One of Poeppel's frequent collaborators, Oded Ghitza, a former Bell Labs scientist and now a biomedical engineer at Boston University, worked with auditory neuroscientist Steven Greenberg to play with the rate of sentences. Consistently across languages, normal conversation runs at about four to six hertz. Ghitza and Greenberg compressed sentences so that what originally took four seconds to speak now took two. Listeners could still understand them. But when they compressed the sentence again, by a factor of three, it got much harder. Poeppel plays me an example and it sounds like a garbled mouthful: badazump.

But Ghitza and Greenberg weren't done. "They take the compressed waveform," says Poeppel, "which sounds for shit, and they just slash it into little slices and they add silence in between." Now it sounds like "bit . . . tut . . . but . . . tit." Poeppel explains what I'm hearing: "The acoustic signal is very small and not informative, totally smeared and gross, but you've added a little bit of empty time in there. You've squished everything and now you insert little silent intervals." Then Ghitza and Greenberg measured how many mistakes listeners made when trying to repeat the sentences. With the speech totally compressed, the error rate was 50 percent. As they increased the intervals of silence, the error rate went down by 30 percent. But then it got worse again. "That's weird, right?" says Poeppel. "There's this moment where you do better, even though the acoustic signal is still totally crappy." That sweet spot came with silent gaps of eighty milliseconds. What Ghitza and Greenberg concluded was that from a brain's-eye view, a listener needs two things. The first is what they term an "acoustic glimpse"—the pitch, loudness, and timbre carried on the sound wave that tell the system to listen in. Then the brain needs what they call a "cortical glimpse," which is simply time—enough time to decode the information that has just arrived.

Poeppel has one last hall-of-mirrors trick he wants to show me. It's one he often uses in presentations. First, he sets a black arrow moving from left to right across his computer screen. "We have this very powerful intuition that time flows like an arrow in this particular direction," he says. "That of course is true. It's the second law of thermodynamics. But at a local level, that turns out not to be quite right." Now he clicks on a recording of a sample sentence. It sounds like high-pitched chirping. The second sentence, too, is high-pitched and choppy, but now I can make it out: "Ants carry the seeds so better be sure that there are no anthills nearby." I can also understand the third, which is similarly distorted: "We've all been rich and spoiled long enough to hate the machine age." (I notice there's never just a sentence about making spaghetti for dinner.)

"One you didn't get, the others you did," says Poeppel. "Why are those of any interest?"

He pauses before delivering the punch line.

"They're all played backward."

"Even the ones we can hear?" I ask, amazed. I knew about the Beatles' famous backmasking on "Revolution 9," but it didn't sound like this.

"That is the rub. Here's a spectrogram of a regular forward sentence." He shows me a now familiar image of splotchy bands of energy. "Now you take a regular forward sentence and you slice it up into little chunks again. But now you flip them backward."

He makes this sound simple, and for auditory engineers it is. Once they have a representation of a sound signal on a computer, they can manipulate it every which way. For a layperson like me, it takes a while to get my head around the idea.

"Flip the whole sentence?" I finally ask.

"No, you flip the slices."

If you flipped the whole sentence, he explains, it would sound syllabic but unintelligible. Each small slice is another story.

"Do the backward slices equate to a word?"

"That's the question. What's the size?"

After playing around with the size of the slices, the researchers who dreamed this up found that the fundamental size that had to go forward was the length of a syllable, or about one hundred fifty to two hundred milliseconds. "When you go to the size of a syllable, you're not allowed to play this game," says Poeppel. "That's one of the reasons the syllable is very foundational to perceptual parsing."

Taken together, what all these acoustic manipulations have demonstrated is that the sounds we hear have to be good enough. "'Good enough' can mean a reasonable amount of spectral information and a clear amount of temporal information," says Poeppel. "You need the temporal information to segment into some kind of unit, and you need the spectral information to begin to initiate the process of looking stuff up. But that's all: If it's good enough, it gets you there." And it starts the cascade of responses rippling through your brain.

17

SUCCESS!

In April 2006, a group of two- and three-year-olds clustered around our dining room table, propped on their knees in their chairs and leaning forward, Spider-Man party hats askew, so as to get closer to the cake.

"Happy birthday to you, happy birthday to you . . ."

Alex had turned three. The deadline that Simon Parisier had set had arrived. We had gotten the cochlear implant in under the wire and already it was making a difference.

Nonetheless, the day was bittersweet.

The boys and girls celebrating with us mostly lived nearby in Brooklyn. As one- and two-year-olds, they had attended the same local child care center as Alex. They were a reminder of how much had changed in our lives.

I tried not to focus on the children's chatter as they ate their cake, but it was inescapable. They were a talkative group.

"For my birthday, I got Spider-Man, too!" said one little boy. "He makes webs."

"I'm going shopping with my mommy after this," said one precocious girl.

"I love chocolate cake!" cried another.

Alex had blown out his candles with enthusiasm and climbed on the living room furniture with some of the boys. When you're two or three, giggling is a universal language. But that was largely the extent of his communication with these children. When he did talk, I was usually the only one who understood him.

In the coming year, I didn't know if Alex would see any of these kids at all. Up to this point, in order to preserve neighborhood friendships and maintain some normalcy, we had kept him enrolled in the familiar child care center near home—Jake and Matthew and some of their closest friends had been there, too. We'd been dropping Alex off there for the afternoons after his morning sessions at Clarke. As a three-year-old, however, Alex would be in Clarke's preschool program all day and would leave the neighborhood behind. To make that work, I was even going to let him take the school bus, an idea that had appalled me when my friend Karen first mentioned it back at the very beginning of this journey. All my certainties about what was possible for a child Alex's age had been knocked upside down.

Now I was (pretty) sure this was the right thing to do. At Clarke, Alex would have teachers of the deaf in the classroom who could do the explicit language teaching he needed, his speech therapy would be a routine part of his day, there would be audiologists on hand, and everyone in the building knew how to manage his equipment. We were fortunate to have such a school available to us.

But the decision marked the first complete diversion from the life we had imagined for Alex, from the life his brothers were leading, a life of growing up in our brownstone Brooklyn neighborhood, where I rarely got to the grocery store without seeing an acquaintance and where the boys rarely went to the playground without meeting a friend, where I could walk them to school every morning and wave to the same set of crossing guards for years. Instead, Alex would be across the East River at the other end of the city from me, with children I hardly knew. Although I would visit occasionally, my sense of his days would be mostly limited to the weekly reports from his teachers and therapists and to the yellow bus that picked him up at seven each morning before his brothers were even awake and dropped him off in the afternoons. Alex was zigging while everyone else was zagging. For a year? For two? Forever? All we could do was work and watch and wait.

"Cat," said the speech therapist. Alex moved a toy car a few inches along the table to show that he heard a short vowel sound.

"Cake." Alex moved the car three times farther for a long vowel.

It was June. Six months into having a cochlear implant, Alex was being evaluated again. The therapist recorded some of what he said, mostly simple phrases like "going to work" and "it broken," but also an impressive five-word string—"Let me have yellow hat"—while playing with Mr. Potato Head, who featured regularly in evaluations. Still, omissions, distortions, and substitutions like "boo" for "blue" and "wowuh" for "flower" littered his speech, leaving him with the articulation of a child a year younger.

"Mama, want mik, pees," Alex called to me in August as he ran through the kitchen with his friend Nina.

"Of course, sweetheart, I'll get you some milk," I answered and got up from my stool to go to the refrigerator.

"Lydia!" Across from me, my friend Amy, Nina's mother, who had known Alex all his life, was looking at me with surprise.

"I understood him!" she said. "That's the first time I've ever understood him." Only now could she admit to me that she never had before.

After a doctor's appointment one morning that fall, I delivered Alex to Clarke later than usual.

"Alex, I've been looking all over for you!" said his teacher when we reached the classroom. She pretended to look behind the door.

"Here I am!" he announced. "I was looking for you, too!" Then he pretended to look under the table.

To my amazement, my sweet but silent baby was turning into a funny boy, quick with a laugh and eager to play along with a joke.

In December, I sat once again at the observation window at Clarke, watching Alex and his classmates play. He had arranged a group of blocks to form a boat and was sitting amidships. When some of his classmates climbed in with him, he handed over a long foam block.

"Row the boat!" he commanded.

Then he picked up another block and pretended it was a fishing rod.

"I have a big fish! Help me pull!"

The children laughed and began to row and pull.

Each sign of progress was thrilling but brought a new dilemma. Almost as soon as we decided to put Alex at Clarke full-time, we had to consider what to do the following year, when he would be four. The goal of a specialized school like Clarke is to prepare children for the mainstream, not to hang on to them longer than necessary. How long was necessary? Would I know by watching through the window? Educating Alex, I was learning, was going to be a moving target. Every year, we would have to reassess and rethink.

Jake and Matthew were in third grade and kindergarten respectively at Berkeley Carroll, a wonderful school five blocks from our home. If Alex was going to join them there the following year, the application would be due in December, long before it seemed we could know if he'd be ready.

I asked the teachers at Clarke to fill out the necessary recommendation forms anyway. When I got them back, I was surprised to see how little indicated that Alex was different from any other child. Of course, his classroom teacher had noted he was behind on every aspect of speech and language. But on a long list of social and behavioral skills such as attention span, ability to focus, curiosity, enthusiasm for new challenges, interaction with peers and adults, Alex looked like a very skillful little boy. Many categories were marked as an "area of strength." Suddenly, I was seeing my child afresh. I had been so fixated on nurturing the seedlings of Alex's language that sometimes all I saw was the dirt, the weeds, and the tender but fragile sprouts of words and sentences. I needed to stand up and look at the whole garden that was my child. Focused on what I feared he couldn't do, I didn't always take time to savor what he could.

Encouraged, I told myself we were just keeping our options open by applying to a mainstream school. On the day the application was due, my truer emotions spilled out as I walked up the street to hand-deliver the forms. My heart started thumping and my palms got sweaty. Half a block from the school, I stopped and stood on the corner for a few minutes. Simultaneously thrilled and terrified, weighed down and released, I wanted to protect Alex and I wanted to push him. I wanted him in a safe haven and I wanted him in the wider world. I looked at the big white envelope in my hands. Once I turned it in, at least, part of the decision would be in someone else's hands.

While we waited to hear from Berkeley Carroll, the anniversary of the activation of Alex's implant arrived in January, a few months before his fourth birthday. It was time to measure his progress. We went through the now familiar barrage of tests: flip charts of pictures to check his vocabulary ("point to the horse"), games in which Alex had to follow instructions ("put the purple arms on Mr. Potato Head"), exercises in which he had to repeat sentences or describe pictures. The speech pathologist would assess his understanding, his intelligibility, his general language development.

To avoid prolonging the suspense, the therapist who did the testing calculated his scores for me before we left the office and scribbled them on a yellow Post-it note. First, she wrote the raw scores, which didn't mean anything to me. Underneath, she put the percentiles: where Alex fell compared to his same-age peers. These were the scores that had been so stubbornly dismal the year before: eighth percentile for receptive language (what he could understand) and sixth percentile for expressive language (what he could say).

Now, after one year of using the cochlear implant, the change was almost unbelievable. His expressive language had risen to the sixty-third percentile and his receptive language to the eigthty-eighth percentile. He was actually above age level on some measures. And that was compared to hearing children.

I stared at the Post-it note and then at the therapist.

"Oh my god!" was all I could say. I picked Alex up and hugged him tight.

"You did it," I said.

From the evaluation, we went straight to Clarke so Alex could go back to school. The director's office was next door to his classroom. I knocked on the door frame and Teresa Boemio looked up from her desk.

"How did it go?" she asked as soon as she saw me. She knew where we'd been.

"Pretty well."

I handed over the Post-it note.

"Wow!" she cried. "WOW!"

She pulled his teacher out of the classroom and showed her the numbers. Then they grabbed his speech therapist. All of us started hugging and whooping as if our team had just won the Super Bowl.

"I don't know if I've ever seen a kid do this well this fast," said Teresa.

There was no way to know exactly how Alex was hearing the

world, but the trajectory of his language development said that whatever he was hearing was good enough. It had made him a success story. Unquestionably, he had a lot going for him: usable residual hearing, teachers and therapists who knew how to help him, parents who talked and read to him a lot, even two older brothers who discussed the finer details of superheroes and how to play hide-and-seek with him day in and day out. But hearing aids alone had not been enough for Alex. The transformation came with the cochlear implant. Like the difference between candles and fluorescents, Alex's residual hearing registered the world dimly and imprecisely except within a few small circles of light, while his cochlear implant lit everything up brightly if unnaturally. Once he had seen everything clearly, he could make better use of the warmth and ambience he got from his natural hearing, and he knew what lay in the shadows. Language followed.

That night, after the family celebrated, Alex sat on his bed and we talked some more about the day.

"You did so well today," I said and did the family dance of success that Mark had invented. It involved circling my fists and hips and chanting, "Go, Alex. Go, Alex."

He laughed.

"So no more processor?" he asked, referring to the external part of his cochlear implant.

I stopped dancing and stared at him.

Had we really never explained that he would always be deaf? It had never occurred to me that he might think otherwise. Every night, when he took off his aid and processor, he went back to being almost entirely unable to hear. After all, the internal implant only worked when the external implant was in place. To tell him I loved him when I tucked him in, I had to get up close to his left ear and shout.

He and we had managed a major achievement, but already we were entering a new phase. It wasn't just education that would be a moving target. His sense of self, his emotions, his identity would shift and evolve. I had had nearly two years to come to terms with Alex's hearing by then. Now it would be his turn. We would have to use his newly acquired language to help him understand. The pride and acceptance of Deaf culture suddenly took on new significance. I found myself dearly hoping that Alex would follow their lead and come to view his deafness simply as a difference. My next job would be to help him do that.

At least his success made it easier to decide to send him to Berkeley Carroll as a four-year-old the next September. I tried not to think about what lay ahead, even though I could guess. Jake was entering fourth grade and reading chapter books, writing short essays, and working with fractions. Matthew was going into first grade and reading as well. Would the sound Alex was getting allow him to accomplish all of that, too? Would he be able to build on the language he had and keep pace?

As I readied the boys for the first day on which they would all attend the same school, however, I thought not about the future but about the effort that had gone into getting us this far. I helped Alex strap on his new pint-size blue backpack, with his name stitched in white, then took his hand to maneuver our stoop while Jake and Matthew waited on the sidewalk. Watching the three of them race to the corner together, our ordinary walk to school suddenly seemed extraordinary.

18

THE SEARCH FOR EVIDENCE

In the early days of cochlear implants, a dramatic jump like the one Alex made was far beyond the imagination of most researchers. In 1987, when Caitlin Parton was implanted, she had a 5 percent chance of carrying on a normal conversation without lip-reading. By 1995, however, the wildest claims about cochlear implants didn't seem quite so wild. The conclusion from that year's National Institutes of Health conference stated that "a majority of those individuals with the latest speech processors will score above 80% correct on high-context sentences, even without visual cues."

The reason was technological. Using Don Eddington's Ineraid device, a far more effective speech processing program had been developed by engineers Blake Wilson and Charles Finley at Research Triangle Institute in North Carolina. Called continuous interleaved sampling (CIS), it used a number of signal processing techniques, some of which had been tried before but never in combination. Instead of extracting features of speech as the Australians had done, this program included all the frequency information present in spoken language and tracked the moving energy in the sounds. It avoided overstimulation by alternating electrodes—hence the term "interleaving." "The basic idea," explains Eddington, "is stimulating one electrode for a short period of time and then turning it off and stimulating another electrode for a very short period of time so that no electrodes are ever on at exactly the same time." The signal was also transmitted at relatively high rates of nine hundred to a thousand pulses per second. "It cycles through like many fingers along a piano keyboard," explains Arizona State researcher Michael Dorman, who collaborated with Wilson. "The amplitude of the pulses mimics the amplitude of the energy. . . . The faster the pulse rate, the more detail [it conveys.]" What CIS achieved was an approach that "allowed the brain of the user to become a far more active and important part of perception."

The difference was plain to see from the start. "There are only a few times in a career in science when you get goose bumps," Dorman once wrote. That's what happened to him when his patient Max Kennedy, who had been participating in clinical trials for the Ineraid, traveled with Dorman to North Carolina to try out the new program for Wilson and Finley. Kennedy was being run through the usual set of word and sentence recognition tests. "Max's responses [kept] coming up correct," remembered Dorman. "Near the end of the test, everyone in the room was staring at the monitor, wondering if Max was going to get 100 percent correct on a difficult test of consonant identification. He came close, and at the end of the test, Max sat back, slapped the table in front of him, and said loudly, 'Hot damn, I want to take this one home with me.'"

Ultimately, he did. Although every company has developed their own particulars, all cochlear implants today use processing programs based in large part on CIS.

Even with better speech processing, however, the sound implants provide isn't much like natural hearing. "It's a fascinating plasticity question," says Elissa Newport, the Georgetown professor who studies language acquisition and miniature languages. "The information you get out of a cochlear implant is very different than the information you get from hearing, and so the brain either has to really adapt in some way that we don't understand, or it doesn't totally but it does somewhat. It's really a very big adjustment that the brain needs to make in order to acquire language from the kind of input signals that a cochlear implant produces."

For Newport, the question is of more than just professional interest, since her husband, Ted Supalla, is deaf and a prominent ASL scholar. She worries about deaf children who don't succeed with implants—"the ones who don't show up on *Good Morning America*"—and, as a scientist, is skeptical of talk of miracles. "I don't believe in miracles; I want to know what the mechanism is."

One lesson from Newport's own research, however, is just how resilient children are when it comes to learning language. She first brought this up while we were discussing sign language and the poor ASL skills of most hearing parents. Her studies have found that parents didn't have to be all that good at signing for the child to benefit—results later echoed in her work on miniature languages. "When you're a kid," says Newport, "your input doesn't have to be perfect, perfectly regular, or perfectly fluent in order for you to become very fluent in your language. Inconsistencies, imperfections in the input, don't seem to deter kids when they're young."

"Isn't that also an argument for cochlear implants?" I ask.

Yes and no, she says. "That doesn't mean you can throw anything at the brain and it'll figure out how the patterns work."

Don Eddington was concerned about this very thing in the early stages of his work. "One worried a lot about making kids worse off in a sense," he says. "The impact of those very weird signals on brain organization was a big question."

To find out what that impact really was, I went to see Mario Svirsky, an electrical engineer at New York University. His interest in the problem originated at the opposite end of the spectrum from

Newport—he was fascinated with the technology. As a teenager in Uruguay, Svirsky was captivated by a science-fiction article that imagined artificial vision, and he decided then that he wanted to help make such a project a reality. Upon arriving in the United States for graduate school, he found no one was really working on artificial vision (though they are now). After a brief flirtation with neural prostheses for paraplegics, Svirsky ended up studying cochlear implants. Despite his engineering bent, he has thought philosophically about questions like Newport's. A slight man with a trim beard and an earring, Svirsky has a gentle, considered air about him, which served him well as he waded into the thorny controversies surrounding implants. He has collaborated with the outspoken anti-implant psychologist Harlan Lane, and Svirsky's copy of Lane's book *When the Mind Hears* bears the inscription "To Mario, Whose Mind Hears." Says Svirsky, "We prodded each other's thinking."

Among other things, Svirsky studies the brain's adaptation process to the degraded signal provided by implants. This work does not answer Newport's question about children directly—by necessity, the work on adaptation is done with adults who were deafened after they had language, because they have something with which to compare the implant signal—but it provides useful information. "The signal is degraded in specific ways," Svirsky explains. "One is spectral degradation, meaning you have fewer channels of independent information." This is essentially the problem presented by having fewer keys on the piano with which to make the same music. "The other [problem] is frequency shift, meaning that the whole signal can be shifted to higher or lower frequencies or distorted in different ways." In this case, the middle C on your piano might sound like A or like high C. The cochlear implant was designed to mimic and thereby take advantage of the organization of the human cochlea, with different pitches lined up like piano keys. Electrodes that are closer to the apex of the cochlea are associated with lower frequencies, those near the base with higher frequencies. Or that's the idea. "Other than the fact that both maps are tonotopic, there are no guarantees that they are a perfect match or even a good match," says Svirsky.

Fortunately, most people's brains do seem able to adapt with practice the way I did to the distortions Poeppel played for me. There are cases, though, where the listener simply cannot make sense of the new signal. For them, Svirsky is creating an approach to fitting implants in which the recipient can fiddle with dials in real time to get the signal to the optimum level for the user's particular understanding. Turning the dial changes the assigned frequencies of the electrodes. The results are not always logical. One woman understood speech better when her electrodes began at higher frequencies, essentially leaving out the lower levels. "If you talk to a phonetician, they will tell you that by getting rid of the frequencies below seven hundred hertz, you're eliminating a lot of very useful information for speech perception," says Svirsky. "However, she did better with this because [it] required a lower level of adaptation."

Like Poeppel, Svirsky has some favorite examples of degraded signals that are particularly related to cochlear implants. The best one includes a visual representation of the problem. In a set of four portraits of Abraham Lincoln, the first is untouched and represents "normal." The second is a heavily pixelated image that is degraded spectrally, with fewer channels of information provided. "You can take an image, get rid of most of the information—this has maybe a thousandth of the original—and yet it's recognizable to any American." Then Svirsky plays me the paired audio recording, which sounds like the computer-altered voice of a kidnapper demanding a ransom. Even so, I can hear that the kidnapper is reciting the Gettysburg Address: "Four score and seven years ago our fathers . . ." The next picture of Lincoln has been frequency-shifted and it has a slightly swirly, hallucinatory aspect but is still clearly Honest Abe. Now it sounds like Donald Duck is doing the reciting: ". . . Brought forth upon this nation . . ." In fact, Donald Duck comes up a lot when new implant users try to describe what they're hearing. The fourth and final image has been manipulated both ways; it is degraded spectrally and its frequencies have been shifted. The picture no longer looks much like Lincoln at all but like a series of black and gray smudges. The accompanying recording is unintelligible, at least the first time through. "You get the sense that this may be something that you might be able to learn, given enough time," says Svirsky. "But the learning process would not be trivial."

"Are you suggesting that some cochlear implant users are hearing the world like that?" I ask.

"Like that, perhaps worse," he says. Then, seeing my dismay, he adds, "It gets easier."

I have to remind myself that even though the kidnapper and Donald Duck delivered very poor renditions of the Gettysburg Address, I heard and understood it.

By the time I started looking for them, the small group of children who had been in the first clinical trials of cochlear implants were adults in their twenties. I found some obvious success stories. Caitlin Parton graduated from the University of Chicago and went on to law school with the hope of working on disability rights issues. Matt Fiedor, who received a single-channel implant from Bill House, did remarkably well with that more limited device. He even majored in Japanese! Mark Leekoff, who had been jeered at the *Jeopardy!* taping, told me he struggled socially when he was young, but at Tufts University he found both a group of friends and academic success. I met him at West Virginia School of Medicine, where he is the first deaf student ever enrolled. The school has been creative about solving some of the issues that have cropped up. They let Leekoff use an iPad app with his stethoscope, since listening to a heartbeat is still beyond the capabilities of his implant. For his surgical rotation, they experiment with clear surgical masks, since Leekoff understands best by combining his implant with reading lips.

But there are stories on the other side as well. Implants didn't work for everyone. I met Peter Hauser at Rochester Institute of Technology, where he heads the Deaf Studies Laboratory, to talk about his work as a neuropsychologist. He was willing to tell me his own story, too. Now in his forties, with an animated face and an easy laugh, Hauser lost his hearing at five from meningitis. As a teenager, he participated in not one but two clinical trials. His prospects were mixed, since he'd had hearing for several years but had also been deaf for quite a while. In 1983, he got a single-channel implant. "Every syllable sounded like a beep," he tells me through an interpreter. "So baseball was two beeps, but the beeps could be loud, quiet, high, low. In the soundproof booth at the audiologist's office, when only one sound was given at a time, I could identify the beep fairly well, the rhythm of the beeps. But when they opened the door, it was all over. I couldn't discriminate the background sounds from the beeps or anything." In the end, that implant was too distracting and Hauser gave up after a year. A few years later, he had the first device replaced with a multichannel implant. "With the twenty-two channel, it was like twenty-two different beeps," he says. He used it for about nine months and, again, gave up. Instead, Hauser, like many others, found a home and a career in Deaf culture when he went to Gallaudet for graduate school.

Individual experiences can't be relied on to tell the whole story, of course, or even much of it. If you know one child with a cochlear implant, I've heard it said, you know one child with a cochlear implant. Even though children have been receiving implants for more than twenty years now, there is still enormous variability in outcomes. On the other hand, as children get implants at younger and younger ages and get better support from audiologists and speech pathologists, there are far fewer really poor performers than there once were. "The difference is for the little ones," says Don Goldberg, an expert in auditory rehabilitation at the Cleveland Clinic and president of the AG Bell Association. "We've seen an impressive change."

The hope was always that implants would improve access to education. The consensus among researchers is clear: Most deaf children benefit from a cochlear implant. "Kids with implants are doing better on average than kids without implants," wrote Marc Marschark, an expert on deaf education at the National Technical Institute for the Deaf, one of nine colleges at Rochester Institute of Technology. However, he adds, implants do not transform deaf children into hearing children. "[They] still generally perform behind their hearing peers."

In an important study published in 2000, Mario Svirsky and his colleagues at Indiana University, where he worked in the well-respected cochlear implant program before moving to NYU, showed exactly that. The researchers examined not just whether a child's language development improved after receiving a cochlear implant but also whether it improved *more* than it would have without an implant. "Language development" is an all-encompassing measure that goes beyond speech production (or intelligibility) and speech perception to include more sophisticated skills such as reading comprehension and grammar. In a normally developing child, chronological age and "language age" rise in lockstep, so that at one year of age a child has a language age of one year, at three years old he has a language age of three, and so on. A graph depicting the maturation of language will show a straight diagonal line from the bottom left corner stretching to the top right. Historically, profoundly deaf children developed language at half the rate of hearing children. "Without an implant, these children would be expected [in English] to speak like a two-year-old when they are four and so on," explains Svirsky. So their language development sheared off from their hearing peers from the start, and the gap only widened over time. "What I found was once they received an implant," he says, "their development started proceeding at a normal rate on average." In other words, the gap stopped widening and the line of language development for a child with a cochlear implant now paralleled the diagonal for a typically hearing child.

In another study, Svirsky showed that the vast majority of children with cochlear implants eventually reach 80 percent or better on open-set tests of speech production. He tested this by taping kids saying a set of standard sentences and playing those recordings to a panel of "naive listeners" who had no experience listening to the speech of the deaf. In such a test, a typically hearing child will score 80 percent at the age of four, 90 percent at five, and then go higher. In Svirsky's study, children implanted in the first two years of life all achieved a score of 80 percent or better by the age of eight. Those who didn't get their implants until the age of three did not do as well.

It was a logical conclusion that age mattered and that the earlier a child received an implant, the less ground she would have to make up. A 2010 study headed by Dr. John Niparko, who directed the cochlear implant program at Johns Hopkins University at the time (he is now at the University of Southern California), showed why earlier was better. It measured spoken language development in 188 children at six centers around the country. As with Svirsky's study, the children did better on spoken language development than they otherwise would have but did not catch their hearing peers. Those who got implants under eighteen months had "significantly higher rates of comprehension and expression" than those who were implanted between eighteen and thirty-six months, though not everyone succeeded to the same degree, and that second group did better than those implanted after thirty-six months. But Niparko found that several other things also helped kids do well, particularly strong parental involvement, higher socioeconomic status, and greater residual hearing.

Back in 2003, some of those factors showed up in an influential study by Ann Geers and colleagues at the Central Institute for the

Deaf in St. Louis. Geers studied 181 eight- and nine-year-old children who received implants between ages two and five. After testing comprehension, verbal reasoning, narrative ability, and spontaneous language production, either in speech or sign depending on the child's preferred language mode, Geers found that some of the same factors that predict success in hearing children were helping deaf children, too—greater nonverbal intelligence, smaller family size, higher socioeconomic status, and female gender all boosted performance. She also found that those who were educated in oral classrooms had better language development than those who weren't.

In 2008, Geers released a study of 112 of the same students, who were then in high school. It is one of very few studies to track students over such a long time, because researchers are still waiting for most implanted children to grow up. "Performance of these students far exceeds expectations for children in previous generations," wrote Geers and colleagues. But academic difficulties remained. Speech recognition and intelligibility got worse for most students in noisy conditions like those of a classroom. Gaps in IQ scores—both verbal and nonverbal—persisted between deaf and hearing children. Most implanted children made consistent progress in reading that paralleled hearing peers, but 20 percent made minimal progress in the eight years between studies, staying at the "fourth-grade barrier." And writing, including spelling and expository skills, was difficult for most of the implanted children; only 38 percent scored even close to the hearing students. Early implantation, parental involvement, higher socioeconomic status, and oral education all helped those who'd done best.

Who your parents are, when you get your implant, how you're educated—these are clearly determining factors of success. Yet there is still surprising variability in performance even for kids who start out in similar situations. According to David Pisoni, a cognitive neuroscientist at Indiana University who is particularly interested in why that variability persists, roughly 15 to 20 percent of cochlear implant recipients will not do well with the device. Nor have cochlear implants erased the problems of deaf education.

Despite a concerted effort to come up with educational solutions over the past few decades and reams of academic papers on a wide variety of classroom-related subjects, there has been distressingly little measurable progress, as if deaf academic achievement is a brick wall with no path over, around, or through. "The only thing we do know is that the median reading level of deaf eighteen-year-olds in the US has not changed in forty years," Marc Marschark told me in 2012. As I had found back when I was searching the Internet, that median reading level, the middle ground, is stuck in the fourth grade.

At least we've moved past the restrictions of the oral-only era. As Andrew Solomon put it in his *New York Times* article, "The insistence on teaching English only . . . served not to raise deaf literacy, but to lower it. Forbidding sign turned children not toward spoken English, but away from language." In that article, Solomon quoted a deaf woman named Jackie Roth: "We felt retarded," Roth said. "Everything depended on one completely boring skill, and we were all bad at it. Some bright kids who didn't have that talent just became dropouts. . . . We spent two weeks learning to say 'guillotine' and that was what we learned about the French Revolution. Then you go out and say 'guillotine' to someone with your deaf voice, and they haven't the slightest idea what you're talking about—usually they can't tell what you're trying to pronounce when you say 'Coke' at McDonald's."

Such frustrations led students of Jackie Roth's all-oral era—who are in their fifties or older today—to become leaders of the Deaf culture movement. But the younger generation that protested at Gallaudet in 2006 was educated differently, because the philosophy underlying deaf education had changed. Even before the Americans with Disabilities Act became law, separate was no longer deemed equal for any child with special needs, and a push for mainstreaming began in 1975, when Congress passed Public Law 94-142, a landmark piece of legislation that guaranteed free, appropriate public education for all children with disabilities. That law was amended in 1986 and again in 1990 with the Individuals with Disabilities Education Act (IDEA). By 1986, only three out of ten deaf children still attended specialized schools like the residential ones that predominated in earlier times. Most of the rest were in public school classrooms of one sort or another, either in separate classes for deaf students or in mainstream classes with interpreters or resource teachers. But they still weren't succeeding. A 1988 Federal Commission, assigned to investigate deaf education, declared in its report: "The present status of education for persons who are deaf in the United States is unsatisfactory. Unacceptably so. This is [our] primary and inescapable conclusion."

By then, many students were getting at least some of their education in ASL. Beginning in the 1970s, deaf educators adopted "total communication." The idea was that deaf students should learn through whatever means possible—spoken English, ASL, fingerspelling, writing, anything that worked. It was a wary compromise that seemed like a good idea in theory, but in practice this either/or approach left students fluent in neither English nor ASL, satisfying no one. The signing of many teachers in such classrooms was often what the deaf call "shouting"—throwing in a sign for a prominent noun or verb here or there. Or it was SimCom, for "simultaneous communication," which requires teachers to speak and sign at the same time—a difficult trick to pull off because the syntax of ASL and English are not the same. Or, worst of all to proponents of ASL, there was Signed Exact English, which is not ASL at all. On the other hand, from the point of view of parents interested in oral education, some total communication classrooms seemed like very quiet places without enough speech, where good oral role models were in short supply.

For a time, a system called Cued Speech seemed promising. A set of hand signs flashed near the face to indicate phonemes, it was expressly designed to help with literacy but has worked less well with English than with other languages.

A new approach, known as bilingual-bicultural, emerged in the 1990s. In such schools, face-to-face interaction is in ASL; English is taught for the purposes of reading and writing. "Bi-bi" is popular in the Deaf world and Gallaudet has recently declared itself a bilingual-bicultural university. Here, too, success depends at least in part on the quality of the signing. The question is no longer "about whether deaf children can be appropriately educated in sign (at least within educated circles)," wrote Marc Marschark and Peter Hauser, but about "the subtle and not-so-subtle implications of varying degrees of sign language fluency."

When I first read Marschark's 2007 book, *Raising and Educating a Deaf Child*, it was a vast relief. It seemed remarkably fair and balanced, mostly by being willing to speak bluntly to both sides of the deaf education debate. In the first chapter, Marschark tackled two common beliefs that hung around like persistent storm clouds. "First, there has never been any real evidence that learning to sign interferes with deaf children's learning to speak," he wrote. "Second, there is no evidence that deaf children with cochlear implants will find themselves, as some had warned, 'stuck between two worlds,' (hearing and deaf) and not fully a member of either." His message was that deaf children need full exposure to a language—any language—from their earliest years and that parents need to be the ones providing it. "Effective parent-child communication early on is easily the best single predictor of success in virtually all areas of deaf children's development."

Marschark, who is hearing, began his career as a cognitive psychologist interested in language and metaphor. As a side project, he got interested in the way deaf kids used figurative language. Intrigued by the results, he retrained himself in child development and the issues of deaf children. Part of that effort involved reading all of the existing research literature, which he compiled into his first book, *Psychological Development of Deaf Children*, published in 1993.

On the strength of that book, Marschark was recruited by the National Technical Institute for the Deaf. NTID was a daunting prospect for someone who didn't really know any deaf people or any sign language, but Marschark has been there for two decades and is now proficient in ASL. And yet, he has written, "as a hearing person I can never truly understand what it means to be Deaf or to grow up (deaf or hearing) in the Deaf community. I may be welcomed, and I may know more about deafness than many other people, but I still have to understand Deaf people and the Deaf community from my hearing perspective."

That was just one of the things we talked about when I spent the afternoon in his office in Rochester. Bearded and bespectacled, Marschark looks like an academic. But he's an academic on a mission. He speaks regularly around the country. Lately, he's been giving a lot of presentations that all have the same theme: What We Know, What We Don't Know, and What We Only Think We Know About ____. Fill in the blank with whatever subject Marschark's been asked to tackle: cochlear implants, ASL, literacy, school placement, you name it. He's managed to offend both sides, including his colleagues. "I warn everybody at the outset: Each of you is going to be upset. But I say, Don't leave! Don't get too upset, because in five minutes you'll be happy and the person next to you will be upset."

This series of talks is based on an epiphany Marschark had after a major research project in 2009. "I discovered that a lot of what we do in deaf education doesn't have any evidence to support it, or people are cherry-picking," he said. Since then, he has been on a crusade to force people to question the "religious" convictions they have of what they think they know is true. Take for example that question of whether sign language gets in the way of learning English. People regularly send him e-mails with lists of research citations ostensibly proving him wrong when he says there's no evidence that sign language interferes. But if you read the studies, he says, "what the research says is that if you want a child to be oral, if you want them to speak, they need to be exposed to spoken language." That is not the same thing as saying that sign language poses a problem, nor is there any established threshold of how much oral experience is enough. "The data's not wrong, but the conclusion is drawn from one side of the coin," he says. Yes, it's true that children in oral programs tend to speak better than children who are not in oral programs but, Marschark points out, "they're not randomly assigned." Children who are likely to do well in oral programs—for instance, those like Alex with some residual hearing—go into those programs. Children who are not likely to do well tend not to go into those programs or to drop out of them along the way.

On the other hand, Marschark, who has long been a proponent of ASL for deaf kids, has dared to say lately that much of the argument for the benefits of bilingualism is based on emotion more than evidence. When we met, he was about to publish a paper that "was going to make people mad." It argued that, so far, no one had shown convincingly that bilingual programs were working. "There is zero published evidence, no matter what anybody tells you, that it makes anybody fluent in either language."

In the earlier edition of *Raising and Educating a Deaf Child*, published in 1997, Marschark was circumspect about cochlear implants. In the 2007 version, he supports them. What changed? "There wasn't any evidence and now there is," he told me. "There are things I believed three years ago that I don't believe. It doesn't mean anybody's wrong; it just means we're learning more. That's what science is about." One of Marschark's favorite lines is: "Don't believe everything you read, even if I wrote it."

"So," I ask him, "what do we know?"

What we know, he says, is that it's all about the brain. "Deaf children are not hearing children who can't hear; there are subtle cognitive differences between the two groups," he says, differences that develop based on experience. Deaf and hard-of-hearing babies quickly learn to pay attention to the visual world—the facial movements of their caregivers, their gestures, and the direction of their gaze. "It's unclear how that visual learning proceeds, but their visual processing skills develop differently than for hearing babies, for whom sights and sounds are connected."

Marschark and his NTID colleague Peter Hauser have compiled two books on the subject, one called *How Deaf Children Learn* and a more academic collection, *Deaf Cognition*. The newly recognized differences affect areas like visual attention, memory, and executive function, the umbrella term for the cognitive control we exert on our own brains. Marschark and Hauser stress these are differences, not deficiencies. Understanding those differences, they argue, just may tell us what we need to do to finally improve the results of deaf education. "We have to consider the interactions between experience, language, and learning," they write.

Such differences would seem to be obviously true of deaf children with deaf parents—and they are. Neuroscientist Daphne Bavelier worked for years with Helen Neville before heading her own lab at the University of Rochester. Bavelier, together with Peter Hauser, who was a postdoctoral fellow in her lab, has continued Neville's work on visual attention. Among other things, she has shown that deaf children have a better memory for visuospatial information and that they attend better to the periphery and can shift attention more quickly than hearing students. The ability is probably adaptive, as it helps them notice possible sources of danger, other individuals who seek their attention, and images or events in the environment that lead to incidental learning. This skill may have a downside, however, because it also makes students more distractible.

Bavelier and others study deaf children of deaf parents, since those kids receive little or no sound stimulation and they learn ASL from birth. Such a "pure" population makes for cleaner research as it limits the variables that could affect results, and it could be said to show what's possible in an optimal sign language environment. But it also limits the relevance of such work. "Deaf of deaf" are a minority within a minority. The other 95 percent of deaf and hard-of-hearing children are more like Alex—born to hearing parents and falling somewhere along a continuum in terms of sound and language exposure, depending on when a hearing loss is identified, what technology (if any) is used, and how their parents choose to communicate. One thing that no one can control is who your parents are.

Children with cochlear implants are another minority within a minority—though they may not be a minority for long. Research into their cognitive differences has barely begun, but what there is sits squarely at the intersection of technology, neuroscience, and education. "No part of the brain, even for sensory systems like vision and hearing, ever functions in isolation without multiple connections and linkages to other parts of the brain and nervous system," notes David Pisoni, a contributor to *Deaf Cognition*. "Deaf kids with cochlear implants are a unique population that allows us to study brain plasticity and reorganization after a period of auditory deprivation and language delay." He believes differences in cognitive processing may offer new explanations for why some children do so much better with implants than others.

To that end, Pisoni and his colleagues designed a series of studies not to show how accurately a child hears or what percentage of sentences he can repeat correctly, but to try to assess the underlying processes he used to get there. In one set of tests, researchers gave kids lists of digits. The ability to repeat a set of numbers in the same order it's heard—1, 2, 3, 4 or 3, 7, 13, 17, for example—relates to phonological processing ability and what psychologists call rehearsal mechanisms. Doing it backward is thought to reflect executive function skills. Even when they succeeded, which they usually did, children with cochlear implants were three times slower than hearing children in recalling the numbers. Looking for the source of the differences in performance, Pisoni zeroed in on the speed of the verbal process in working memory, essentially the inner voice making notes on the brain's scratch pad. From these and other studies, he concluded that some brain reorganization has already taken place before implantation and that basic information processing skills account for who does well and who doesn't with an implant. Those with automated phonological processing and strong cognitive control are more likely to do better.

For his part, Marschark is particularly interested these days in "language comprehension," a student's ability to understand what he hears or reads and to recognize when he doesn't get it. Marschark tested this by asking deaf college students to repeat one-sentence questions to one another from Trivial Pursuit cards. Those who communicated orally understood just under half of the time. Those who signed got only 63 percent right. All were encouraged to ask for clarification if necessary, but they rarely did so. Whether they were overly confident or overly shy, the consequences are the same. These kids are missing one-third to one-half of what is said. You can't fill in gaps or ask for clarification unless you know the gaps and misunderstandings exist; and you have to be brave enough to speak up if you do know you missed something.

Hauser has been exploring the connection between language fluency and the development of executive function skills. "Executive function is responsible for the coordination of one's linguistic, cognitive, and social skills," he says. The fact that many deaf children show delays in age-appropriate language means they may also be delayed in executive function. Too much structure or overprotectiveness—something parents of deaf kids are prone to—compounds the problem by further stifling the development of executive function skills. So far in the study, deaf of deaf are on par with hearing children, "so being deaf itself isn't causing the delay," Hauser says. When we met, he was starting to collect data on deaf children with hearing parents.

If researchers can continue to pin down such cognitive differences, says Marschark, they might be able to pass that useful knowledge along to teachers. "Bottle them and teach them in teacher training programs," he suggests. "Here's how you can offset the weaknesses, here's how you can build on the strengths."

"That would be great," I acknowledge.

I'd also been struck, however, by how much keeping up with research on deaf education feels like following a breaking news story on the Internet. It leaps in seemingly conflicting directions simultaneously. Furthermore, not all of it feels relevant to Alex.

"Absolutely," Marschark agrees. "Kids today are not the kids of twenty years ago or ten years ago or even five years ago. Science changes, education changes, and—most important—the kids change. And one of the problems is that we as teachers do not change to keep up with them."

The one constant, he points out, is that they are all still deaf, and no one—not deaf of deaf or children with cochlear implants or anyone in between—should be considered immune from these cognitive differences. Those with implants, he says, will miss some information, misunderstand some, and depend on vision to a greater extent than hearing children. (The latter is a good thing because using vision and hearing together has been shown to consistently improve performance.) As Marschark and Hauser asked rhetorically in *Deaf Cognition*: "Are there any deaf children for whom language is not an issue?"

19

A PARTS LIST OF THE MIND

—

Davīd Poeppel is pulling books off the groaning shelves in his office at New York University. I've come back to see him to talk about language, the other half of his work.

"Open any of my textbooks," he says, holding one up. "Why is it that any chapter or image says there are two blobs in the brain and that's supposed to be our neurological understanding of a very complex neurological, cognitive, and perceptual function?"

What he is objecting to is the stubborn persistence of a particular model depicting how language works in the brain. "We're talking about Broca's area and Wernicke's area," he says. "It was a very, very influential model, one of the most influential ever. But notice . . . It's from 1885." He laughs ruefully. "It's just embarrassing."

In the 1800s, the major debate among those who thought about the brain was whether functions were localized in the brain or whether the brain was "essentially one big mush," as Poeppel puts it. The evidence for the mush theory came from a French physiologist named Jean Pierre Flourens. "He kept slicing off little pieces of chicken brain and the chicken still did chicken-type stuff, so he said it doesn't matter," says Poeppel.

Then along came poor Phineas Gage. In 1848, Gage was supervising a crew cutting a railroad bed in Vermont. An explosion caused a tamping iron to burst through his left cheek, into his brain, and out through the top of his skull. Amazingly, Gage survived and became one of the most famous case studies in neuroscience. He was blinded in the left eye, but something else happened, too.

"Here's a case where a very specific part of the brain was damaged," explains Poeppel. "He changed from a churchgoing, very Republicanesque kind of guy to essentially a frat boy: a capricious, hypersexual person. Why is that result important? Because it's about functional localization. It really said when you do something to the brain, you affect the mind in a particular way, not in an all-out way. And that remains one of the main things in neuroscience and neuropsychology. People want to know where stuff is."

It was a few years after Gage's accident that Pierre Paul Broca, a French physician who wanted to know where stuff was, or more precisely whether it mattered, met his own famous case study. The patient, whose name was Monsieur Leborgne but who was known as Tan, had progressively lost the ability to speak until the only word he could say was "tan." Yet he could still comprehend language. Tan died shortly after Broca first examined him. After performing an autopsy and discovering a lesion on the left side of the brain, Broca concluded that the affected area was responsible for speech production, thereby becoming the first scientist to clearly label the function of a particular part of the brain. A few years later, German physician Carl Wernicke found a lesion in a different area when he studied the brain of a man who could speak but made no sense. The conclusion was that this man couldn't comprehend the world because of his lesion and therefore that that area governed speech perception. From these two patients—one with an output disorder, the other with an input disorder—a model was born.

"The idea is intuitively very pleasing," says Poeppel. "What do we know about communication? We say things and we hear things. So there must be a production chunk of the brain and a comprehension chunk of the brain."

Poeppel turns to his computer screen.

"Here's vision."

He pulls up a map of a brain's visual system—actually a macaque monkey's visual system, which is very similar to that of a human. Multicolored, multilayered, a jumble of boxes and interconnecting lines, the map is nearly as complex as a wiring diagram for a silicon chip.

"Here's hearing."

Up pops a map of the auditory system. A little less complicated, it nonetheless has no fewer than a dozen stages and several layers and calls to mind a map of the New York City subway.

"And here's language."

Up comes the familiar image, the same one that was in so many of the offending textbooks he has pulled off the shelf, showing the left hemisphere of the brain with a circle toward the front marked as Broca's area and a circle toward the back marked as Wernicke's area, with some shading in between and above.

"Really?" says Poeppel. "You think language is reducible to just production and perception. That seems wildly optimistic. It's not plausible."

The old model persisted because those who thought about neurology didn't usually talk to those who thought about the nooks and crannies of language and how it worked. A reason to have that conversation—to go to your colleague two buildings down, as Poeppel puts it—is that perhaps we can look to the brain to learn something about how language works, but it's also possible that we can use language to learn something about how the brain works.

Poeppel's goal is to create "an inventory of the mind." His approach to studying language is summed up in a photograph he likes to show at conferences of a dismantled car with all its components neatly lined up on the floor. "What we really have to do is think like a bunch of guys in a garage," he says. "Our job is to take the thing apart and figure out: What are the parts?"

He began by coming up with a new model. Poeppel and his colleague Greg Hickok folded together all that had been learned in the previous hundred years through old-fashioned studies of deficits and lesions and through newfangled imaging research. "I say to you 'cat.' What happens?" asks Poeppel. "We now know. . . . You begin by analyzing the sound you heard, then you extract from that something about speech attributes, you translate it into a kind of code, you recognize the word by looking it up in the dictionary in your head, you do a little brain networking and put things together, you say the word."

First published in 2000 and then refined, their model applies an existing idea about brain organization to language. It is now widely accepted that in both vision and hearing, the brain divides the labor required to make sense of incoming information into two streams that flow through separate networks. Because those networks flow along the back and belly of the brain, they are called dorsal and ventral streams. Imagine routing the electrical current in a house through both the basement and the attic, with one circuit powering the lights and the other the appliances. In vision, the lower, ventral stream handles the details of shape and color necessary for object recognition and is therefore known as the "what" pathway. The upper, dorsal stream is the "where/how" pathway, which helps us find objects in space and guides our movement. In hearing, there's less agreement on the role of the two streams, but one argument is that they concern identifying sounds versus locating them.

Hickok and Poeppel's model suggests that this same basic dual-stream principle also governs neuronal responses for language and may even be a basic rule of brain physiology. "There are two things the brain needs to do to represent speech information in the brain,"

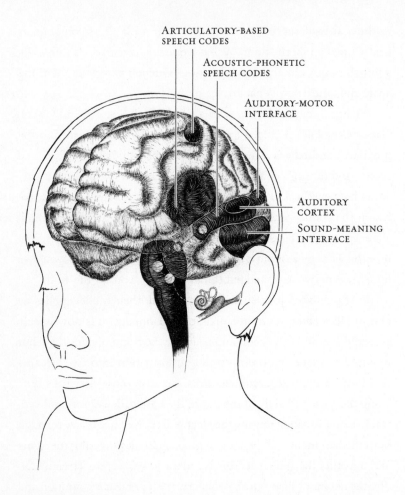

ARTICULATORY-BASED
SPEECH CODES

ACOUSTIC-PHONETIC
SPEECH CODES

AUDITORY-MOTOR
INTERFACE

AUDITORY
CORTEX

SOUND-MEANING
INTERFACE

Hickok explains. "It has to understand the words it's hearing and has to be able to reproduce them. Those are two different computational paths." In their view, the ventral stream is where sound is mapped onto meaning. The dorsal stream, running along the top of the brain, handles articulation or motor processes. Graphically, they represent their model as seven boxes connected by arrows moving in both directions, since information feeds forward and backward. Labeling each box a network or interface—articulatory, sensorimotor, phonological, sound analysis, lexical, combinatorial, and conceptual—

emphasizes the interconnectedness of the processing, the fact that these are circuits we're talking about, not single locations with single functions. Some change jobs over time. The spot where sensory information combines with motor processing helps children learn language but is also the area where adults maintain the ability to learn new vocabulary. That same area requires constant stimulation to do its job: It's not only children who use hearing to develop language; adults who lose their hearing eventually suffer a decline in the clarity of their speech if they do not use hearing aids or cochlear implants.

The model also challenges the conventional belief that language processing is concentrated on the left side of the brain. "You start reading and thinking about this more and you think: Can that be right?" says Poeppel. When he and Hickok began looking at images from PET and fMRI, they saw a very different pattern from lesion data alone. "The textbook is telling me I'm supposed to find this blob over here, but every time I look I find the blobs on both sides. What's up with that?" says Poeppel. "Now it's actually the textbook model that the comprehension system is absolutely bilateral. It's the production system that seems more lateralized. You have to look more fine-grained at which computations, which operations, what exactly about the language system is lateralized."

It was a bold step for Poeppel and Hickok to take on such ingrained thinking. Although it initially came in for attack, their model has gained stature as pretty much the best idea going. "At least it's the most cited," says Poeppel. "It may be wrong. . . . No, it's absolutely for sure wrong, because how could it possibly not be way more complicated than a bunch of colored boxes?" But at the moment, the model has achieved the enviable position of being the one up-and-coming neuroscientists learn and then try to take apart. I even found it in a new textbook.

Notice that this model does not just locate "language," which, despite how we often talk about it, is not one monolithic thing. Nor does

it limit itself to perception and production. Instead, the model's organization parallels the organization of language, which consists of elements like sound, meaning, and grammar that work together to create
what we know as English or Chinese. Each of those linguistic tasks, or
subsystems, turns out to involve a different network of neurons.

Phonology is the sound structure of language, though ASL is now
thought to have a phonology, too, one that consists of handshapes,
movements, and orientation. "During the first year of life, what you
acquire is your phonetic repertoire," says Poeppel, referring to the
forty phonemes often identified in English. "If I'm English, I have this
many vowels; if I'm Swedish, I have nearly twice as many; that is my
inventory to work with—no more, no less. That's purely experience-
dependent, because you don't know what language you're going to
grow up with." One thing Poeppel and Hickok observed was that the
ability to comprehend words and the ability to perform more basic
tasks like identifying phonemes or recognizing rhyming patterns
seemed to be separate. This was an important observation because it
meant that phonemic discrimination might not be required to understand the meaning of words. Their model puts the phonological network roughly where Wernicke's area is, in the posterior portion of the
superior temporal gyrus. As for Broca's area, which is part of the inferior frontal gyrus, they and many others now think of it as a region
that handles as many as twelve different language-related processes.

Morphology refers to the fact that words have internal structure—
that "unbelievable" is made up of "un-" and "believe" and "-able," and
maybe "believe" has some parts as well. It turns out there's a heated
academic debate between psychologists and linguists about whether
morphology even exists—"excruciatingly boring for nonexperts, but
interesting for those of us who nerd out about these things," says
Poeppel. Morphology is likely processed, says Poeppel, "in the inter-
play between the temporal and inferior frontal regions."

The same is true of syntax, the grammar of language—structure at the level of a sentence. Here there is no argument: You have to have sentence structure for language to make sense. It's what tells us who did what to whom, whether man bites dog or dog bites man.

Semantics concerns the meaning of words, individually and in context in a sentence. They are not the same. Even if you know the different meanings of "easy" and "eager," the brain has to do some extra work to make sense of the difference between the following two sentences where the structure is seemingly identical:

John is easy to please.

John is eager to please.

They differ by exactly one syllable, by one tiny sound sequence, but they mean something entirely different for John. He's either the pleaser or the pleasee. A fluent user of language understands that effortlessly. Lexical ability, too, is located in the ventral stream, specifically in the posterior middle temporal gyrus and the posterior inferior temporal sulcus.

Prosody is the rhythm of speech, the contour of our intonations, and to Poeppel its importance is clear. "How is it that you can distinguish subtle inflections at the end of the sentence and know that it's a question?" says Poeppel. "Syntax doesn't tell you that. This small change in acoustics changes the interpretation completely." It is also a cue for prominence, according to Janet Werker, a developmental psychologist at the University of British Columbia and colleague of Athena Vouloumanos. In seven-month-old bilingual babies, Werker found that the infants were using prosody (pitch, duration, and loudness) to solve the challenging task of learning both English and Japanese, languages that do not follow the same word order ("eat an apple" in English versus "apple eat" in Japanese). Babies cleverly listen for where the stress is placed in a sentence. Prosody, it seems, plays a part in how you break the sound stream into usable units. That process begins in utero, where a baby can't hear all the specific sounds of his mother's voice but can follow the rhythm of her language. Its location in the brain is similar to that of phonology.

Discourse refers to units of language larger than a sentence. A favorite of the deconstructionists of the 1980s, it's a relatively new field of study that looks at writing, conversation, etc. It is discourse, I realize, that is the aim of a literate, educated person. "The assumption is that you get that for free," says Poeppel, meaning that if you learn all the more basic parts of language, you will achieve discourse as well. I'm not so sure about that, since one of the emerging areas of concern for cochlear implanted children is how their language develops beyond elementary school, when it needs to take on sophistication and subtlety as they read to learn and begin to write essays. In the brain, discourse would fall into what Poeppel and Hickok call the widely distributed "conceptual network," meaning we bring to bear much of our neural resources to tackle it.

For all of these language systems, the scientific focus has been on mapping, on figuring out "where stuff is." But that is not enough, says Poeppel. "We have more blobs, better blobs. But our yearning should be bigger. Namely, what's an explanation for what's going on there? What is the mechanism? A mechanism is some kind of account or explanation for how a set of elements interacts to generate something. That's not what we have."

In the search for a mechanism, linguists and neuroscientists are asking if there is a hierarchy to these subsystems of language. They are searching for what Poeppel calls the "primitives," the primary colors of language with which everything else is constructed. For my purposes, thinking about language, and thinking about Alex, in terms of this new list of parts, I see how each of them contributes to the next. Phonology is an obvious problem for a child who doesn't hear well, but morphology figured in many of Alex's progress reports. If he couldn't hear or say parts of words that had meaning, like the "-ed" on a past-tense verb or the "s" of a plural, he wouldn't always understand or be understood. Semantics showed up in the constant effort to expand his vocabulary. Prosody had previously been drowned out for me by the drumbeat of concern over perception and production, but I could see how helpful it is because of how integral it is to the ability to make sense of the stream of language. "One of the absolutely elemental things you have to do for comprehending language is you have to segment it," says Poeppel. "If you can't segment it, you can't actually look up the words at all. If you can't look up the words, your syntax isn't going to be all that great, either."

"So how can we use this information to help a child?" I want to know, even though I recognize that Poeppel spends his time in the lab, not the classroom.

"The syllable has practical ramifications," he says. "From the get-go, what we would want to give a kid are these cues in the signal. The syllable is a segmentation cue that provides rhythmic information and that's easy to remember. That's super-useful. The information comes prepackaged." And the brain apparently makes use of that fact. Various studies have shown that the brain gives a bit of extra effort to syllables that occur at the beginning of words, something researchers call the "word onset effect." Poeppel has also shown that the brain resets itself to track the syllable rate of a speaker.

How can we help children train that ability and lay down those circuits? One way is to let them explore the rhythms of language. Happily, that is something that comes naturally or that was instinctually built in to many children's stories and songs. Usha Goswami, an educational neuroscientist at Cambridge University, studies whether dyslexia may be related to difficulties in sound processing. Her research suggests that a return to some old-fashioned wordplay would be useful long before children learn to read. "Nursery rhymes are perfect little metrical poems," she says. "We know children love them, and they enjoy things like singing in time to music." Engaging in such

activities seems to help develop the necessary phonological and pro-
sodic networks. Her laboratory is also studying the benefits of having
five-year-olds learn poetry by heart.

Goswami even got me thinking about the neurobiology of *The
Cat in the Hat*. "Repetition matters, too," she says. "That's true of all
learning by the brain, the more repetition the better. Dr. Seuss hasn't
just got rhyme, he's also got repeating phrases. The child who's good
with syllables will be quicker to get onset rhymes and will be quicker
to acquire phonemes once they acquire abstract units like letters.
That's why developmentally you want to start with amplifying the
syllable level by bringing in all this rhyme. That should put the child
in a better position to begin reading."

As soon as I got home, of course, I read to Alex about the cat who
came to play on that cold, cold, wet day.

20

A ROAD MAP
OF PLASTICITY

In one corner of a classroom at a Head Start facility in Eugene, Oregon, three preschool children sit at a small table trying to color a frog green without going outside of the lines. It's not easy for them and they have to concentrate. In the opposite corner, another group of children are playing with balloons, batting them back and forth and trying to keep them aloft. These three, with the balloons, have been assigned the role of Dr. Distractor, and their job is to try to get the children at the table to stop paying attention to their work. Over the next eight weeks, every Tuesday night, the two groups will swap roles regularly, and whoever is in the distracting group will move closer and closer to the table until, by the last session, they are practically on top of those doing the drawing. To maintain concentration, the poor four-year-olds at the table have to marshal all their resources, looking only at their papers, thinking hard about their work. At one point, a little girl raises her hand to physically hide the other children and their balloons from her view.

The entire process is an exercise in applied neuroplasticity and it represents a very modern way of thinking about how to help children.

The old way can be summed up in a story that Mike Merzenich, who made his name studying plasticity, told me. In the 1960s, around the time that David Hubel and Torsten Wiesel were conducting the research on kittens that would establish the concept of the critical period, one of Merzenich's relatives, who taught elementary school in Wisconsin, was honored as the national teacher of the year. The family gathered to celebrate her accomplishment and she gave a small speech. "She stands up and says, 'My secret was that I figured out, on the basis of testing, what children were really worth my attention and I gave them everything.'" Merzenich was so bothered by that statement that he's never forgotten it. "What do you think the kids that weren't worth her attention got?" he asks. "Nothing."

The story is proof, for Merzenich, of why it's so important to understand exactly how the brain changes with experience. "[Hubel and Wiesel] did fantastic science, and the description of the critical period was brilliant," he says, but he argues that the cast on the interpretation—the idea that the window of development slammed shut so firmly—was wrong. "It had very destructive negative consequences. What it meant in American society—and world society—was when you came to the schoolhouse door, what you saw was pretty much what you had. Kids were in a sense doomed to their genetic fate. It was imagined by pediatricians and child psychologists and schoolteachers everywhere that your brain was fixed and it was all about compensation for the circumstances in front of you."

Merzenich's groundbreaking work with monkeys in the 1970s, at the same time as he was contributing to the invention of the cochlear implant, showed that brains can and do continue to change throughout life. "It's a revolution," he exclaims, "understanding that in a sense we are continuously remodelable, that in fact, our fate and our ability are under our direction and charge. To understand that we can at any point improve, bodily improve, the things we do in life and that this science can be applied to change the outcomes of children on a large scale, this is revolutionary. It will come into every aspect of human societies—pretty soon it will be everywhere."

One place the understanding has already arrived is the classroom. The effort to take what we know about brain development and make use of it in schools has been christened "neuroeducation." From studies of executive function in toddlers to research on math ability in adolescence, all of the work takes as its foundational principles that learning is driven by brain circuitry and that brain circuitry is wired by experience.

The transition out of the laboratory is not always smooth. Not every basic scientist wants to wade into the classroom or the home and apply what has been learned. Those who do sometimes seem to be tripping over one another in the race to create scientifically based curriculums and video games (Merzenich was one of the first to market such a product, a reading intervention program called Fast ForWord). For their part, educators are not always happy to see the scientists. The two groups don't always speak the same language or share the same immediate goals; a school principal focused on seeing test scores go up will not want to wait for a double-blind study from which some children, by design, won't benefit. Furthermore, educators have grown suspicious of perennial promises of the next big thing in learning as well as defensive over complaints about the job they are doing. Nevertheless, it's undeniable that researchers today do know far more than they used to about how children learn and how their brains develop.

To talk about some of the knowledge underpinning this new trend and see how it might apply to Alex, I turned to Helen Neville at the University of Oregon and the program she and her colleagues have been working on with a local Head Start for nearly a decade. Given that Neville set out to change the world back in the 1960s, it's not surprising that she thinks it's incumbent on scientists to try to use their work to improve children's lives. "Now we're seeing where the rubber meets the road," she likes to say.

"It all starts with basic research," she tells me. "You can't jump in and pull an intervention out of the air. You have to know what systems show neuroplasticity, what ones are vulnerable, how fixed they are, how changeable they are, and what are their mechanisms, so you can target them. I've been studying this for thirty years. I've kind of come full circle, I think."

The connection between preschoolers trying to pay attention and deaf adults exhibiting better vision in the periphery may not be immediately obvious, but there is actually a straightforward plotline to Neville's story of scientific discovery, and her areas of focus have a special resonance for me. She began by studying the brains of deaf and blind people because their experiences were so unlike those of hearing and sighted people. "You had to start with a population that would give you a good chance of finding a difference," she says. From there, she moved to people who had had slightly different experiences, such as second language learners or those who learned sign language instead of spoken language. And finally, she looked at typically developing children who have different experiences "just by virtue of being different ages or in different stages of cognitive development."

All of the work is ongoing, but after thirty years Neville has sketched out a nuanced picture of how the brain's systems of vision, audition, language, and attention change with experience. As Mike Merzenich maintains, the window does not necessarily shut firmly all at once. But neither, emphasizes Neville, does it stay open indefinitely for every skill. The details of what she calls "profiles in plasticity" matter. Some brain systems are so hardwired that they are the same whether a person hears or sees or not. Some change considerably with experience, but only during particular windows in development, which are in turn determined by the specific system in question— hearing has one, language turns out to have several. And then there are certain parts of the brain that continue to be shaped by experience throughout life, "with impunity," says Neville. Knowing all of this has allowed her to chart "a road map of plasticity."

It was a map I thought would be useful to have in hand for what it could tell me about the remaining uncharted territory of language learning. Elissa Newport had given me a glimpse of how the brain approaches learning language. David Poeppel laid out a plan for how language works once we're good at it. Helen Neville was going to connect the dots.

No matter how many languages a child learns before the age of seven, they will all operate in the brain in a similar fashion once they are acquired. Additional languages learned later are processed differently, and Neville tries in part to pinpoint those differences. Work from her laboratory and others has demonstrated that there are different profiles in plasticity for phonology, syntax, and semantics. As a result, scientists have created something like an evolving account of the development of language in the brain, allowing us to follow along as a child's linguistic capabilities mature. The definition of an adult response can vary, depending on the skill—some responses become smaller and more efficient over time, others more widespread. What matters is the change in the way neurons do their work, a sign that they are wiring together with use or being pruned from disuse.

It won't surprise anyone who has mangled the sounds of a second language—that is, most of us who didn't start to learn until high school—to know that the window for phonology, the ability to recognize speech sound contrasts, opens and closes early. Generally, we will have an accent in any language we learn past the age of seven. As babies, we are already working on the underlying skill, learning to distinguish the sounds of our native language—the ones we'll need—from sounds we don't need. This is the process by which English babies learn the difference between "r" and "l," but before the end of their first year, Japanese babies let that distinction fade, since Japanese doesn't contain those sounds. Furthermore, the better babies are at responding to the contrasts of their native language at the ripe old age of seven and a half months, the more proficient they are at language as toddlers: Their word production and sentence complexity at twenty-four months are higher and so is the mean length of their utterances at thirty months. In Neville's laboratory, they found that in thirteen-month-olds the brain response to known words differs from that to unknown words and is broadly distributed over both hemispheres. By twenty months, the effect is limited to the left hemisphere, more as it is in adults, reflecting increased specialization, vocabulary size, and maturity.

The ability to chop up language into usable chunks—to segment it—is also different in late learners. They can do it, of course, or they wouldn't be able to make any sense of the new language, but they do it much more slowly. The "word onset effect," part of the N1 that comes early in native speakers, isn't as visible in late learners.

Starting early with language doesn't just get rid of accents, it also makes it much easier to handle grammar and produce the appropriate sentence structure fluently. We acquire those syntactic skills nearly as early as we master phonology. An early study by Elissa Newport shows this starkly. In English-Korean speakers who came to the United States at varying ages, scores on a test of English grammar drop off steeply between the ages of seven and seventeen. At two, children's brief sentences already reflect the word order of their native language, so those learning English know that "Daddy" comes before "eats" and "pizza" comes after. (This is the very skill that Janet Werker attributed to prosody.) Between the ages of three and five, children are already experts at such grammatical fine points as how to describe something that happened in the past, how to ask a question, how to say that you don't want or like something. They hit these milestones no matter what language they are learning, which, as Neville has pointed out, "supports the proposal that language learning has a significant biological basis."

Differences don't just show up in second language learners. How proficient you are in your native language is evident in your brain as well. It's a rather infamous fact that much cognitive research is done on college students—if you are a researcher at a university, students are handy and cheap. But are they typical? When he first started reading up on the literature describing how the brain was organized for language, Neville's colleague Eric Pakulak found it fascinating until he looked at the methods section in one particular paper. "It's twenty undergraduates from Harvard," he says. "That's not really representative." As it happens, Pakulak plays rugby with a group of men who do represent a wide swath of the socioeconomic spectrum. So he decided to test a group that included some of his teammates and others who weren't university students. He looked at two particular responses in the brain that reflect syntactic processing (grammar): the early anterior negativity, a wave that falls somewhere between a hundred and three hundred milliseconds and is relatively automatic; and the more controlled P600, the late wave thought to reflect repair and reanalysis. In higher-proficiency speakers, Pakulak found that the early response was more efficient and more concentrated. He hypothesizes that this "frees up" mental resources for the later response, which is larger and more widespread.

For those who miss out on developing the early responses, there are alternatives. Pakulak also studied Germans who spoke English as a second language. Because of the consistency of the German educational system, all had begun learning English around the age of eleven or twelve—late in brain terms—but all were good at it. (This is the university bias at work again: All the subjects were undergraduates, graduate students, or professors.) Pakulak found that the late start on English meant the Germans lost the benefit of the early response, but they made up for it by putting more brain areas to work later in the process. "They're using different resources to achieve a different level of proficiency," he explains. "They're using more controlled processes because they don't necessarily have the access to these early and automatic processes that are more constrained by differences in experience." Those results reminded me of what Greg Hickok told me: that anyone, like Alex, who got less input, was going to have to use more top-down processing. The same apparently was true of German graduate students who wanted to study in the United States.

Here's the good news: We can keep learning new words in any language for as long as we like. "This is a system that continues to change throughout life," says Neville. "It can be set up in a native-like fashion even if you start learning a language at the age of thirteen or fifteen." In Chinese-English bilinguals who started learning English either very young, as preteens, or later, the systems in their brains that handle semantics look "completely identical no matter when English is learned."

No matter what language we're learning or when, it helps to pay attention. "When you focus on something specific, the brain produces a bigger response to it," says Neville. Many changes in vision, audition, and language observed in neuroplasticity studies depend in part on selective attention. (This was true of Merzenich's monkeys, by the way. The changes in cortical areas came only when they paid attention to the spinning disks they were being asked to touch.) I remembered that David Poeppel had said he could tell by glancing at an N1 whether the listener had been paying attention. In fact, the response can be 50 to 100 percent stronger, says Neville. Children's ability to pay attention matures just as their language does, but even at the age of three they are able to listen to one story over another, and when they do, there's a visible change in their brains.

To show me how they test that in the lab, Eric Pakulak puts me in a soundproof booth where there are speakers to my left and right and a television screen in front of me. From the control room, he starts playing both audio and video. In my right ear, I hear the children's story of the Blue Kangaroo. In my left, another children's story, from the Harry the Dog series, is playing. I'm supposed to listen to the Blue Kangaroo story and ignore Harry. On the screen, pictures corresponding to the kangaroo story help me in my efforts to focus. When I zoom in on the story mentally—focusing the flashlight of my attention, as neuroscientists say—I find that I can do it. Harry fades to the background. But every so often, my attention wanders and Harry leaps back into my awareness. I have to work to keep him in his place.

In the actual test, the subjects would be wearing electrode caps for EEG recordings. Superimposed on both stories, there are extra beeps serving as auditory probes. What Pakulak, Neville, and their colleagues are really measuring is the neural response to those probes. They wanted to see if the response was larger when the subjects were paying attention. The answer was yes. In adults, it's about twice as large. Although the response is different in children of six, seven, and eight, there is still a clear enhancement when the children pay attention.

The real-life consequences of paying attention—or not—play out in the classroom. Those who are less focused learn less. It's that simple. Brain circuitry drives learning and it does a better job of it when the networks are working at optimal levels. I am not talking here about attention deficit disorder, which is a problem of a higher order, though obviously one on the same continuum. I'm just talking about the more routine ability to listen to the teacher or keep one's concentration on the page.

There is more than one kind of attention, it turns out, and the various forms also have different profiles in plasticity. Sustained selective attention, the kind that is particularly useful in a classroom, takes a very long time to develop. On the one hand, that explains why it's so hard for small children to stay focused. On the other hand, it means the window of opportunity to help them do it better stays open for longer. Neville has found that children of lower socioeconomic status have trouble with selective attention, particularly in tasks requiring the filtering of irrelevant information.

With so much work under her belt, Neville decided it was time to take what she knew about neuroplasticity and figure out how to use it to help children. "We've reached the point where we've learned an awful lot," Pakulak tells me. "We're taking those results and having them inform our development of intervention and training programs."

They started with attention for several reasons. It qualifies as a clear example of Neville's concern that what is enhanceable in the brain is also vulnerable—strengthened, attention helps; weakened, it's a major problem. The good news was that it seemed that attention might be trainable. "If you have a measure that's a predictor of being at risk for delays, that gives you a lot of information," says Pakulak. Attention training might provide a lot of bang for the buck because its effects would be widely felt. It was also a necessary fundamental skill. "Having intact sensory and cognitive systems is really not worth that much if you do not have control of your emotions or have good social cognition," says Neville. That point informs a wide range of current neuroeducational interventions, which most commonly home in on emotional regulation and executive function.

Pulling together the many pieces of the program took time, but now Neville's group has some intriguing results. One hundred forty-one three- to five-year-old children enrolled in the Head Start program were randomly assigned to three groups: One group simply continued to attend Head Start; a second got attention training during the school day for forty minutes a day, four times a week, with three sessions for parents over eight weeks; and the third section brought children and parents to the facility once a week for eight weeks for the children to receive some form of training and the adults to get proven strategies for parenting.

In addition to Dr. Distractor, the attention training included exercises like "Emotional Bingo," in which kids took cards with the names of emotions printed on them, showed what those emotions look like and how they feel, and then learned techniques for calming down, like taking a big "bird breath" and saying, "Oh, well." Children also worked on focusing attention by playing games that required them to start and stop listening, looking, or moving when teachers cried "Freeze." In parenting sessions, "we assessed the way they used language with children, and we encouraged them to engage in more balanced turn taking," says Pakulak.

Children in the attention classrooms—those who took a ride on the "brain train"—made the biggest gains across the board on standardized measures of language and nonverbal IQ, including working memory. Groups of children whose parents had also received the more extensive weekly training did the best of all. Problem behavior went down in the children and social skills increased. Using EEG, the children all underwent the same study of auditory attention in the lab that I did and tried to focus on the Blue Kangaroo and ignore Harry the Dog. "There were changes in cognition," says Pakulak. "We saw neural enhancement with focused attention within a tenth of a second, even with three- to five-year-olds. In the winning group, we saw greater gains in that measure. It made them look like higher socioeconomic status kids."

Will those gains last? To answer that, Neville's group has been following the children longitudinally, bringing them into the lab and visiting them in school every year. The data from the first eighteen months suggested that the kids were maintaining their gains, but only time will tell.

"Attention is a force multiplier," says Neville. "It's key to everything. If you want to learn soccer, if you want to learn to play the cello, if you want to learn to use your cochlear implant, it doesn't matter. . . . If you don't have the ability to focus and suppress distracting information, you can't do anything. And if you do have control of it, you can do anything."

Straightforward though it was, I thought this might be one of the most useful pieces of information I'd learned yet.

21

"I Can't Talk!"

MOMMY! MOMMY! MOMMY! MOMMY!" the voice was screaming. It was early morning and we were at our farm in central New York State. When I'd gone out a few minutes earlier to walk the dog and enjoy the bright and chilly December day, the boys had all been asleep. Usually, they'd have stayed that way for at least another hour. From the hill behind the house, I turned and looked back. A small figure was standing in the doorway on the front porch, yelling and waving. It looked like Alex.

I turned and ran.

"I can't talk! I can't talk!" he cried, panic-stricken, when I reached him. His face was bright red and tears were streaming down his cheeks. It took me a few minutes to realize what was happening. He could no longer hear himself talk.

After Alex lost the hearing in his right ear completely, we expected he would probably lose the remaining hearing in the left ear, too. Every time he smacked into a table or came too close to a flying toy, my heart skipped a beat. We would frantically do our listening checks, but everything stayed marvelously the same. Eventually, our anxiety level lessened. Two and a half years had passed and he was now five.

But the night before, he had had a bad fall. He was climbing a ladder to the sleeping loft in our cabin when he slipped and fell several feet to the floor, smack on his face. I had been very worried, but all evening there had been no change except the spectacular bruise developing across the bridge of his nose. Now he was hysterical.

There had probably been internal swelling in the night, and the rest of his hearing was finally gone. I thought I'd been prepared for that, but his reaction threw me. Until that point, when Alex took off his processor—the external part of his cochlear implant—and his hearing aid to go to bed or get in the bathtub, we had still been able to communicate in a limited way. I had to get close and yell into his left ear. But he could hear me. Once he had cracked the code of language with the help of the implant, it had gotten easier. Probably by combining speechreading and what he knew of language, he managed well enough for me to help him with his bath.

I had assumed that when he lost the remaining hearing, it would be this limited conversation we would miss. He hadn't told me and I hadn't realized—presumably he hadn't realized either—that for him, hearing his own voice was the all-important distinction between his profoundly deaf right ear and his hard-of-hearing left ear. The hearing in his left ear hadn't been enough to kick the spoken language motor into gear on its own, and he couldn't possibly have heard himself very well, but something was not at all the same as nothing.

I ran for my purse. The night before, after the fall, we'd all gone out to dinner. Perhaps shaken up by the accident, or just worn out from playing all day at the farm, Alex had become sleepy early and crawled into my lap at the restaurant. He handed me his processor and hearing aid and I put them in my bag. They were still there in the morning when he woke up. I pulled them out and put his processor on. He could hear me again. He could hear himself again. I rocked him in my lap and tried to soothe him. He began to calm down, but we were both shaken. Mark was up in the fields doing the early-morning farm chores;

to get him, I would have had to leave Alex, and I wasn't going to do that. Besides, I knew there was nothing he could do. Jake and Matthew, who'd been awoken by Alex's screams, were hovering in the background, unsure what was going on. "What's the matter, Mommy? What's the matter?" they asked anxiously. I tried to explain. They crowded onto my lap and added to our hug.

It was December 30. We were more than two hundred miles from New York City, and several feet of snow lay on the ground. It was only seven in the morning. Dr. Parisier's office wasn't open yet. With three anxious children to keep calm, I pulled out the Dog Bingo game they loved and we played a few rounds. I felt surreally as if I were at the eye of the hurricane, waiting for the wind and rain to whip up again.

When Mark came near enough—to the closest corral to feed the goats—I ran out to explain what had happened.

At nine, I called Dr. Parisier's office.

"How long will it take you to get here?" his receptionist, Eva, asked.

"About four and a half hours. It depends on traffic."

She put me on hold briefly, then returned to the line.

"He was supposed to leave at twelve for the holiday, but he'll wait. Come as fast as you can."

I had Alex dressed and in the car in minutes. We'd decided that I would take him into the city alone and the others would stay at the farm. It seemed quicker and more efficient.

Everyone stood waving in the driveway.

"Hey, buddy," I said as I pulled into the road. "It's not every day we get to go somewhere just you and me. And, wow, you get to pick the movie you want to watch! No arguing with your brothers." Like comfort food, he picked *Elmo Visits the Firehouse*, which the bigger boys thought was entirely beneath them by that point.

The sun was glaring off the snow, and I had to keep reminding myself to slow down on the slippery, winding roads.

It was close to two in the afternoon when we reached Dr. Parisier's office, but he was waiting. The hearing test showed a nearly flat line at the bottom of the chart. Alex couldn't hear anything softer than ninety decibels—a train whistle or a jackhammer. The audiogram of his left ear now looked like his right. "Profound hearing loss in both ears," wrote the audiologist.

The vestibular aqueduct that sat above Alex's inner ear had ruptured. Because he had hit his face rather than the side of his head, it had taken longer for the swelling to do its damage, but the damage was done.

Dr. Parisier prescribed an intense course of steroids, which, if taken quickly enough after the trauma, might stop the inflammation before too much harm had been done. We'd been warned that this was the course of action we'd have to take. That was why I'd driven pell-mell through the frozen countryside to get back to the city.

"How often does it work?" I asked.

Parisier hesitated. "I've seen it work," he replied.

Our chances are not good, I thought.

Back home in Brooklyn, Alex and I went from one pharmacy to another in search of the prescribed steroid. It was dark, getting toward closing time, and no one seemed to have it.

"What's it for?" asked one pharmacist. "Allergies?"

How to answer that?

"My son hit his head. If he doesn't get this medicine today, he could lose what remains of his hearing." I tried to quiet the panic I could hear in my voice. Alex was holding tight to my hand.

"Sorry. I could get it by Monday, after the holiday."

Up the street, the same answer.

"What's it for? Croup?"

After a few minutes of looking on the shelves, no luck. Sensing the desperation on my side of the counter, the pharmacist dug out the

phone book so I could call other pharmacies. I called four. No luck. I
was beginning to be unable to speak a coherent sentence.

There was a metal chair pushed up against a rack of Ace bandages
in the far aisle, where customers sat while waiting for prescriptions to
be filled. I collapsed into it and began sobbing. The people in the store
were glancing toward me and then looking uncomfortably away. I
was embarrassed and horrified that I was so completely losing it in
front of Alex.

And he was equally horrified. "Mommy, don't cry. Mommy, don't
cry. We'll find it." He clambered onto my lap and put his arms around
my neck. He started crying, too. I knew better, oh so much better. I
am utterly and completely blowing it, I thought. There we sat in the
pharmacy's dinky metal chair, mother and son, clinging to each other
and weeping.

A voice interrupted. "I found it."

I looked up. The pharmacist, a young Hispanic woman with long
dark hair, was holding up a box triumphantly. "It was hidden away."

I didn't know what to say. "You have it? You have it!"

Alex exclaimed through his tears, "She has it!"

It had been hidden on a back shelf. We had it. We blew our noses,
wiped our eyes, paid for the steroids, and walked back through the
dark streets—it was now night—to the empty house.

Alex and I had never spent the night in the house alone. I promised him he could sleep in my room. But first, he had to take the medicine.

It tasted awful. I tried to make him just gulp it down. I made it
into a milk shake but diluted it too much, so now he had to drink an
enormous glass of bitter shake. He took tiny sips. I bribed. I cajoled. I
begged. I threatened. I held his nose. I added syrup and anything else
I could think of. I lost it again.

"I'm sorry I fell, Mommy," Alex said in a small voice.

Oh god. We sat on the floor and hugged.

"I wish Dr. Parisier could just make it better," he said.

"I do, too."

Again, I was blowing it. I remembered the mantra of Deaf culture: "There's nothing wrong with Alex." Clearly, I was not sending that message. As far down the path toward profound deafness as we already were, I was surprised by how desperately I was clinging to what hearing remained. I really, really wanted him to take the medicine. Although I'd given the cochlear implant most of the credit for his transformative progress, I understood by now how much help he was getting from his left ear and how useful it would be for learning to read, which he had yet to do. I had also been rattled by Alex's hysteria. I wished he was old enough to tell me more about how he was experiencing the world.

Four hours later, he had downed three-quarters of it. Enough.

When we went to bed, Alex insisted on keeping his processor on his head (and still does).

The next day, New Year's Eve, we drove back to the farm in a blizzard. It took us nearly eight hours to get back. It was four below zero, but Mark was waiting in the driveway.

For the next ten days, Alex did battle with the milk shakes. I let Mark take over making the concoctions. He bribed, cajoled, threatened, and did vaudeville acts. Alex took the medicine.

And it worked.

On the ninth day, back in Brooklyn, we were putting the boys to bed. Mark was doing an unscientific hearing test, one that traditionally made the boys laugh. Standing behind Alex's left ear, with no cochlear implant on the other side, Mark said loudly, "Scooby-Dooby-Doo." And Alex said, "Scooby-Dooby-Doo." Then he turned around and looked at Mark in surprise.

"Daddy, I heard you."

"You heard me?!"

"I heard you!"

"Hamburger."

"Hamburger."

"Banana, banana, banana."

"Banana, banana, banana."

"He heard him!" we all shouted. Mark threw Alex up on his shoulders and we all did a victory dance. But quickly, we lifted him down. Gently.

22

THE READING BRAIN

Ask an average four-year-old to tell you the difference between "cat" and "hat," and he will probably tell you that one is an animal you pet and the other is something you put on your head. Ask the same question when he's five, and you might hear that "cat" starts with a "k" sound and "hat" with a breathy "hu" sound. Something profound has occurred. Once a child understands that there's a relationship between the spoken words "cat" and "hat" that has nothing to do with their meanings and everything to do with how they sound, he is launched on the challenging obstacle course of learning to read. All along the way, he must master successive, compounding skills until he soars to the finish line ready to read fluently, or is left behind mired in the mud.

That first hurdle, phonological awareness, requires recognizing that spoken words are made of parts. For those who've completed the obstacle course, that no longer seems like such a big deal. Good adult readers know it so intuitively and act on it so gracefully that we often don't remember what a leap the idea required. We don't go to school to learn to talk, because spoken language, for the hearing, comes naturally. That doesn't mean it's easy, just that it's mostly unconscious.

"Our brains know that there are separate phonemes and syllables within words," says Ken Pugh, a cognitive psychologist at Haskins Laboratories in New Haven, Connecticut. We don't need to be aware of that fact to acquire speech. "We're wired to talk and listen and chew gum at the same time if you will." Reading, on the other hand, as Steven Pinker has put it, "is an optional accessory that must be painstakingly bolted on." It is also a relatively recent cultural invention that requires a child to actively consider and come to some understanding about how language works. That process begins with phonological awareness.

I had to make the leap all over again when Alex started to learn to read. Until then, I hadn't had to think much about the way that spoken language underpins written language or that the term "phonological awareness" refers to speech rather than print. It is the ability to hear that "big" has one syllable and "bigger" has two, that "big" and "ball" begin with the same sound, that "big" and "pig" rhyme, or finally, at the most sophisticated level, that "big" consists of three sounds: "b-i-g." In my ignorance, I was apparently not alone. "People hear this term all the time, parents in particular, but they still don't quite understand," says Haskins scientist Stephen Frost. "Many educators as well don't get it."

Haskins Laboratories, a nonprofit research institute affiliated with Yale and the University of Connecticut, studies the science of the spoken and written word. It might be considered a church of phonological awareness. Certainly, no group of researchers has done more to further the mission of understanding the connections between spoken language and reading—or in the case of people with reading disabilities or hearing loss, the misconnections. They have shown not only how important phonology is for typically developing children but also that it is the leading cause of reading problems. Their work has such obvious implications outside of the laboratory that Haskins scientists spend a good amount of time spreading the word. Donald

Shankweiler, the pioneering reading researcher from whom I borrowed the lines about the cat and the hat and the four-year-old and the five-year-old, likes to tell his colleagues stories of talking to rooms full of educators and suddenly seeing them begin to nod their heads when they have seen the light.

Shankweiler himself saw the light at Haskins in the late 1960s and early 1970s. When he and his longtime collaborator Isabelle Liberman began to test explanations for reading problems, there was a pretty strongly held view that reading was visual and that reading disability was probably visual as well. "Reversals of letters and words were still considered to be the hallmark of dyslexia," wrote Shankweiler. Treatments included eye exercises and motor patterning. Shankweiler and Liberman approached the subject differently. "Because writing transcribes language, it seemed natural to ask how reading builds on the foundations of the child's development of primary language," wrote Shankweiler. It was especially natural at Haskins, which, under the direction of Isabelle's husband, Alvin Liberman, already specialized in speech science and was a breeding ground for new ideas. Al Liberman regularly walked the halls asking, "Made any discoveries today?"

In order to read an alphabetic language like English or Spanish, children need to master the alphabetic principle that letters correspond to speech sounds or phonemes. That much was clear. A phoneme is the smallest unit of a word that has a distinct sound and, depending on dialect, there are about forty in English, such as the "c" in "cat" or the "sh" at the end of "push." The ancient Greeks are credited with being the first phoneticians and "doggedly perfecting the correspondence between letters and all known sounds," wrote Maryanne Wolf, director of the Center for Reading and Language Research at Tufts University and author of *Proust and the Squid: The Story and Science of the Reading Brain*. The discovery that "the entire speech stream of oral language could be analyzed and systematically segmented into individual sounds," she goes on to note, "is not an obvious perception for anyone, in any era. . . . [Yet] the great breakthrough by the inventors of the Greek alphabet . . . happens unconsciously in the life of every child who learns to read."

At Haskins, the Libermans, Shankweiler, and another scientist named Ignatius Mattingly began to debate how it was that children arrived at the alphabetic principle, since the alphabet did not actually directly correspond to the sounds of speech. They came up with the idea of phonological awareness as a preliminary step. No one had really thought to look for such a thing before. In their earliest studies, they found that many preschool children have made a start on phonology—they are aware of syllables—but very few can separate out a phoneme. In the decades of research that followed at Haskins and elsewhere, it has been clearly established among other things that preschool oral language skills are predictive of fourth-grade comprehension skills, that how the brain reacts to speech sounds on the first day of school has big implications for what that child has to do to learn, and that more specifically, the location of a kindergartner's brain response to sound is correlated with how many words she will read per minute in second and fourth grades. Furthermore, says Usha Goswami, the Cambridge researcher who discussed Dr. Seuss with me, "There's now twenty to thirty years of research across many different languages which shows that individual differences in brains that are good at reading and brains that struggle to read are really related to sound structure."

Ken Pugh has been doing some of that research at Haskins for more than twenty years and is the current president of the laboratories. Amiable and clean-cut, he has the look of a softhearted New York City beat cop from an earlier era. Nice as he is, he has no patience for those who have yet to get the message. "It still continues to be the case that many people teach reading in a non-evidence-based way," he says. "The work here on phonological awareness, that's the evidence." He is not the only one who thinks so. In 2000, a National Reading Panel funded primarily by the National Institute of Child Health and Human Development summarized fifty years of accumulated evidence to arrive at five principles for teaching reading. The first is that phonologic awareness is a precursor. The second is the alphabetic principle, or phonics, "the idea that those units can be represented systematically by squiggles," says Pugh. "You can't build a building without a basement; phonics is the basement." Third is building vocabulary, because you also can't have a good reading system without a good vocabulary. Fourth is strategies for comprehension that kick in as kids go beyond decoding and gain fluency. The fifth and final principle, "under-appreciated but important," says Pugh, is motivation, which encompasses cognitive skills like attention—Helen Neville's "force multiplier"—and working memory and planning, all of which contribute to reading outcomes. This last also takes into account the plain old desire to read, a powerful aid in the process since the relationship between reading and all of these elements is reciprocal: Reading both requires and builds phonological awareness, vocabulary, and background knowledge in ways that mean, as one expert wrote, "the rich get richer and the poor get poorer."

It took the National Reading Panel report to put an end to an unfortunate era known as the reading wars, a decade or two in which supporters of phonics did battle with those who believed in a "whole language" approach. "The idea [behind whole language] is that reading is as natural as anything else and you'll pick up what you need, so don't kill and drill on phonics—let's tell kids to use text to guess what's coming next," says Pugh. "This is fundamentally inconsistent with the data. If kids are typically developing and you put them in whole-language approaches, they're not going to do as well, but they may just squeak by because they'll pick up on it on their own. But if kids are at risk for reading problems and if phonological awareness and that kind of thing doesn't come free, you are essentially creating what [we] call curriculum casualties. It's criminal." The state of California is a case in point. In 1987, it adopted a curriculum that favored whole language over basic decoding skills. By 1993 and 1994, three out of four children in the state were reading below grade-level averages. Soon most schools had switched back to emphasizing letter-sound correspondence, exactly as later recommended by the National Reading Panel. Today, although a few proponents of whole language persist in their cause in the United States (with many more in other countries), the current reading debate centers on how best to teach the principles in the national report and help children master what reading expert Maryanne Wolf calls the three code-cracking capacities: phonological, orthographic, and semantic.

The goal is fluency, an ability to read quickly and understand, and to fall back on decoding only for difficult, unfamiliar words like "pericardium" or "obliterative." Fluency affords an opportunity to go beyond the text, and the luxury of time to think. All of which you need, as Wolf points out, to appreciate when you read *Charlotte's Web* not only what might have happened to Wilbur had Charlotte not intervened but also how sophisticated the spider's reasoning really was.

How best to achieve fluency using the five principles is a discussion that is increasingly informed by neuroscience. Over the past fifteen years, Pugh and others have used the full complement of imaging techniques to better understand how reading circuits are built in the brain, what happens when they go awry, and how intervention and learning can help. "A skilled adult reader can decode words, pull them off the page, in what we estimate to be about 250 milliseconds. That ain't a lot of time," says Pugh. "If you can do that effortlessly and automatically, then you can read sentences and put your energy and your thought into the syntax and the pragmatics and understand the story. But if pulling each word off the page takes your whole soul and a lot of time—this is what happens in dyslexia—then by the time you get to the end of the sentence, you tend to forget the beginning and you end up with what appears to be problems of comprehension. . . . If you can't read the words fast enough, it's hard to have everything in short-term memory so that you can operate on what the story is about."

The gold standard for diagnosing reading problems is nonsense—literally. The same skilled reader who took 250 milliseconds to process a word will take another 250 milliseconds to say it out loud. A nonword like "clart" or "tove," which by definition the reader won't have seen before, will take only an additional fifty milliseconds. "Why does that matter?" asks Pugh. "Because being a skilled reader means that you've developed these mappings between letters and sounds so well that you can essentially decode anything like a machine. And you can decode things you've never seen almost as fast as things you've seen a zillion times." If children are struggling, whether they hear or don't hear, they are slow, labored, and error-prone on real words and are "just horribly challenged by nonsense words," says Pugh.

All of this is visible in the brain. Neural activity shifts and concentrates as children learn. The beginning reader looking at a word will use more of the brain in both hemispheres and from front to back—the occipital lobes that control vision, the temporal and parietal lobes that are essential for language, and the frontal lobe that controls executive processes. Over time, the activity coalesces primarily in the left hemisphere, using less of the frontal lobe (anything automatic requires less thinking) and less of the right hemisphere (because language networks have become more fully engaged). Fluency has a signature as clear as John Hancock's—a concentrated response in the perisylvian cortex, which runs along the temporal, parietal, and frontal axis, home to language processing areas.

One of several recent studies that captured this process of change was done at the University of Oregon and involved eighteen kindergartners in an fMRI scanner. The children watched a series of images flash for less than two seconds at a time. Some were lowercase letters such as "k" or "c." Some were false fonts that looked like a letter but weren't (for example, a "c" flipped to open to the left and squared off just a touch). If the same letter or false font flashed twice in a row, the child had to press a button. Those who were considered at-risk for reading showed no difference in response to letters or false fonts. But those who appeared to be on track showed slightly more brain activity in left-hemisphere areas when looking at letters. The false fonts activated more of the visual system that corresponds with object recognition and fewer language areas. After eight weeks of school and an additional thirty-minute reading intervention daily for the at-risk group, the brains of both sets of children had changed. The response in those who began on track had matured to look more like adult readers. Those who had been at-risk looked more like their peers who'd begun ahead of them.

"Reading is an exercise in plasticity," says Pugh. "It's taking lots of systems in the brain that do different jobs with language, memory, attention, vision, and associative learning and turning them into these really efficient circuits that allow you to get from eye to meaning in a couple of hundred milliseconds." Success, then, depends on the brain's ability to connect and integrate these various areas, each of which matures on its own timetable. That's one reason why children the world over usually begin to learn to read at five or older—until that point, they are not biologically ready. One necessary change is that areas that have been wired for speech adapt to also receive visual information in the form of letters. "What reading demands in hearing children is to get away from vision and into language as quickly as possible, because ultimately you want to use the biologically specialized systems for phonology, syntax, and comprehension," explains Pugh. "At some level your grandmother could predict that, but at another level it's profound."

This does not diminish the importance of vision. We need vision to read print. (In blind readers of braille, similar brain processes are at work from the point at which the information—received by touch—reaches the language areas at about two hundred milliseconds.)

Important work by French neuroscientist Stanislas Dehaene and cognitive psychologist Bruce McCandliss of Vanderbilt University suggests the existence of a visual word form area—a spot toward the back of the brain in the left occipitotemporal lobe that seems to specialize in recognizing text. Dehaene calls it "the brain's letter box." The eyes impose constraints on reading, he points out in his book *Reading in the Brain: The New Science of How We Read*, because they can take in only a little bit of information at a time, usually up to twelve letters. Good readers, though, manage to read four hundred to five hundred words per minute. So from the first step of visual analysis, those readers are rapidly transmitting the incoming visual information to other brain areas. Each system in the brain plays a role and, in working with its neighbors, is changed by the experience. "The brain becomes multimodal and that changes everything," says Pugh. "It's no longer auditory speech or visual. It's relational. It's combinatorial."

Sound and vision together help determine the continuum of difficulty for learning to read in different languages. A transparent language like Italian, the easiest to learn, looks as it sounds. "Every letter maps onto a single phoneme, with virtually no exceptions," notes Dehaene. Mandarin Chinese, the most difficult language, does not. Its thousands of characters usually transcribe whole words and must generally be memorized. Italian children can learn to read in a few months; Chinese children are still working on it well into middle school. English and French children are in the middle. They may not have to master Mandarin, but Dehaene, who argues for transparent spelling, notes that "an immense gap between the way we write and the way we speak [in English] causes years of unnecessary suffering for our children." If you don't believe him, consider "bow" and "bough" and "tow" and "tough."

In any language, a brain that can read is forever altered. Proof of that came from studies of women raised in small Portuguese villages. It was the tradition in those villages that one daughter in a family would be educated and the others would be married off without much schooling. As a result, researchers were presented with a ready-made population in which women of otherwise similar intelligence and background differed only in that one knew how to read and the other did not. In another nonsense-word study, the women listened to progressively harder nonsense words and were asked to say back what they'd just heard. "It's called auditory shadowing," explains Pugh. "You hear, you say. No reading. Women who were literate were better at hearing those sounds and getting them out of their mouths in the right way." He pauses for emphasis. "Think just for a second. Literacy, which has changed the brain, actually has a benefit on speech perception." When the researchers looked at activity in the brains of the women, they found that the literate women were using the reading circuits they had developed in order to do the task.

A second study showing something similar was done by researchers Mark Seidenberg and Michael Tanenhaus back in 1979. In a study of hearing literate adults, they played two words. The subjects had to decide if the words rhymed. If they heard a pair of words like "pie"/"tie," they would be faster to say they rhymed than a pair like "rye"/"tie." Why? Because "pie" and "tie" are spelled the same and "rye" and "tie" are not. Only a reader knows that.

The promise of the new understanding of how reading works in the brain is that it might allow educators to catch problems earlier. As it is, children are often well into elementary school before they're identified as needing help. Dennis and Victoria Molfese are hoping to change that. Both psychologists at the University of Nebraska, they have been working for years on identifying patterns in very young children's brains that predict reading problems later. As early as the 1980s, the Molfeses used EEG to measure infants' brain waves and found that a slower response to acoustic stimuli such as "ba" and "ga" correlated with stronger or weaker language skills and vocabulary size as the children got older. In 2000, Dennis Molfese published a study showing that he could use brain responses to sound in newborns to predict with 80 percent accuracy which children would struggle with reading at age eight. By the year 2011, Molfese said his predictions were up to "about ninety-nine percent" correct.

If this is true, what can we do to change the outcome? That would be easier to answer if we knew exactly what it is that's going wrong in the processing for dyslexic readers. This is the focus of Usha Goswami at Cambridge University. "It may be that we need to think more about the kind of acoustic cues in the signal that give you information about where syllables begin and which syllables rhyme with each other," she says. Her theory about what might be occurring is grounded in work by David Poeppel.

About ten years ago, Poeppel was wrestling with the question of exactly how we chop up what we hear in order to process it efficiently. Trying to reconcile the need to sample the world in phoneme-size chunks—the standard thinking—but also seemingly in slightly longer, syllable-size chunks, Poeppel thought perhaps it was possible to have it both ways. What if, he wondered, the mind listens to the world through two separate time windows, a fast one for phonemes of twenty to fifty milliseconds and a slower one for syllables, more on the order of one hundred fifty to three hundred milliseconds? "One way the brain proceeds is to break complicated scientific problems into many problems," he says. He imagined essentially that as you hear the world—music or speech—your brain records two CDs, one for the left hemisphere and one for the right. But when your brain samples them, a certain asymmetry sets in: For the CD on the left, you look at more of the fast-rate information and less of the slow-rate and vice versa for the CD on the right. Then you combine the two for the fullest possible picture.

Goswami's theory is that dyslexic readers struggle in the slower time window. In particular, they have trouble discerning the sharp increase in signal intensity, the "rise time," that accompanies the start of a new syllable. Another prominent researcher, Anne-Lise Giraud, argues the opposite, that it may be that dyslexic readers struggle with the faster time window that governs phonemes. But either way, the general idea supports Goswami's remedies, which include nursery rhymes, poetry, and music.

All of those suggestions are echoed by other researchers who've looked at interventions that build phonological awareness. Poetry sharpens the developing ability to hear the smallest sounds of language, argues Maryanne Wolf. Old-fashioned children's rhymes such as Mother Goose include alliteration, assonance, rhyme, and repetition, she notes, all of which help the cause. Wolf's group recently published a study showing that kindergartners who got more musical training demonstrated greater phonological awareness than those who got less. In addition, an early, well-known experiment in the United Kingdom demonstrated the power of rhyme to facilitate reading. Lynette Bradley and Peter Bryant worked with four groups of four-year-old children. Two of the groups got special training on words that either started with the same sound (alliteration) or rhymed, and were asked to put together those that shared sounds. One of those groups was also shown a letter that matched the shared sound. When tested several years later, those who had received the rhyme training were much better at phonological awareness and learned to read more easily. If they had also seen the matching letter during training, they did best of all.

There's at least one other striking correlation between early experience and later reading success. "Learning to read begins the first time an infant is held and read a story," wrote Wolf. "How often this happens, or fails to happen, in the first five years of childhood turns out to be one of the best predictors of later reading. . . . As they listen to stories of Babar, Toad, and Curious George and say 'goodnight moon' every evening, children gradually learn that the mysterious notations on the page make words, words make stories, stories teach us all manner of things that make up the known universe."

What if the child can't hear *Goodnight Moon*? What if, as for Alex, bedtime stories are falling on deaf ears? Given all that I had learned, it was no longer surprising that deaf and hard-of-hearing children struggled so much with reading, yet all the more worrisome when Alex's hearing got worse just as he was beginning to learn to tackle books. In addition to the inherent difficulty of acquiring phonologic awareness when you can't hear spoken language, there is the stark fact that for many decades, even centuries, deaf children did not learn *any* language until they entered school. Even today, there are a few children in this situation. I have met some.

There have really been two fundamental changes for deaf and hard-of-hearing children. One is the cochlear implant, the other the introduction of early hearing screenings. Since 1993, the percentage of newborns being screened has risen from 3 to 95 percent. The goal today is to identify children with hearing loss by three months of age and intervene by six months. Such intervention can mean hearing aids, sign language, infant/parent speech therapy sessions in which the baby is exposed to spoken language, or all of the above. At twelve months of age, children who are good candidates can receive one or, more often these days, two cochlear implants. That is a radically different situation from the one in which children were routinely not identified as having a hearing loss until they were three or four years old; and it would seem to bode well for later reading success. Acquiring a complete first language early in childhood—any language—is critical for later reading comprehension, and a child who learns either English or ASL early will develop the same language circuits in the brain. But they do need to learn that first language well.

I also now understood that it wasn't arbitrary that mean deaf reading levels are stuck at fourth grade. That is the moment at which most children either have or have not gained fluency. Those who don't get it by then, who hit what researchers call the "fourth-grade slump," will probably never get through the rest of the reading obstacle course.

But, intriguingly, some deaf children—without cochlear implants—*do* read well. "Now we have a massive mystery," says Ken Pugh, who serves as an advisor for a research center at Gallaudet. "What is phonologic awareness if you don't have sound? It's entirely possible that it still exists. Without the ear, it's going to have to be redefined in some way. Are there different ways to build a reading circuit?" For neuroscientists, educators, and parents of deaf and hard-of-hearing children, that is a fascinating and urgent question. The deaf readers themselves don't seem to know how they do it, just as skilled hearing readers might have a hard time explaining exactly what they're doing as they flip the pages of a gripping book. "One of the challenges for signers is that learning to read English is two things: It's learning to read and it's a new language," says Pugh. Research on bilingualism that's specific to spoken and visual languages might be particularly useful. But there might be other answers as well. When I talked to him, Pugh and his colleagues at Haskins were in the process of working with ASL expert Karen Emmorey, a neuroscientist at San Diego State University, to try to secure a grant for a study that would look at the question of how successful deaf readers achieve fluency.

One possibility would be that deaf students can develop phonological awareness either through years of articulation training—so that they have experience producing phonemes, if not perceiving them—or through fingerspelling, speechreading, the phonology of ASL (handshapes, articulation, etc.), or through a combination of such strategies. Research with students at Gallaudet some time ago did indicate that some may be using phonology in spite of the fact that they can't hear. The studies made use of a phenomenon called priming, in which the word presented first makes you faster or slower at reading the word that follows. It works even better if the prime is presented subliminally. If the target word is "tribe," for instance, you will recognize it faster if I prime you with "bribe" rather than with "blurb." Because English has so much irregular spelling, however, you will be fifty milliseconds slower to recognize "touch" if I prime you with "couch." They are spelled the same, but they sound different. "None of that should matter to deaf readers and yet it does," says Pugh. They sped up for "bribe"/"tribe" and slowed down for "couch"/ "touch"—not to the same degree as hearing readers but enough to show a measurable effect.

On the other hand, many researchers argue that the emphasis on phonology is misplaced for deaf readers. In a recent review, Rachel Mayberry, a professor of linguistics at the University of California, San Diego, analyzed the results of fifty-seven studies testing the relationship between phonological awareness and reading ability in deaf individuals, and found that phonology didn't matter much. Only half of the studies had found statistically significant evidence that the deaf readers were using phonological awareness. Overall, phonological skills accounted for only 11 percent of the difference in reading abilities. A bigger factor seemed to be "language ability," which referred to fluency in either English or sign language, depending on the person's preferred mode of communication. In the seven studies where it was measured, that language skill accounted for 35 percent of the difference in reading abilities.

Mayberry is among those who have shown that deaf children of deaf parents with high proficiency in ASL are stronger readers than deaf children with hearing parents who have delayed exposure to ASL (and no real spoken language). The mean reading level of the group with strong ASL skills was post–high school; the mean reading level of those with late exposure to language—ASL or English—and who had poorer ASL skills was between third and fourth grade. (Individuals with cochlear implants or good early oral language were not included.) That is a dramatic difference, showing clearly that ASL as a first language is not an impediment to literacy and is quite possibly a boon. There is a caveat: The study is correlational, as deaf education expert Marc Marschark noted to me, meaning that strong ASL skills and strong reading skills appear together in this group, but not that one necessarily causes the other. What it does demonstrate is that lack of early language is an enormous problem with long-term implications. Presumably, the deaf-of-deaf children spent a lot of time communicating with their parents in their early years. The other children, who effectively had no language until they reached school, could not possibly have done so. That is why the two clear, inarguable messages of Mayberry's work and many other studies are those that Marschark stresses: Early access to language—any language—and parent-child interaction matter more than anything else.

It's partly for this reason that groups like the American Society for Deaf Children, which has a membership of about three thousand families, advocate for exposure to ASL for all deaf children with whatever auditory input families want added on. The group's president, Beth Benedict, who teaches in the Department of Communication Studies at Gallaudet, told me that the group sees it not as a question of options or choices but of opportunities. "Too many of my students didn't know ASL until much later," she wrote to me. "Too often their audiologists told them not to sign. In fact, many said their spoken and written English improved after learning ASL." She points to research like Mayberry's that shows the benefits of ASL for language development, and says that it also boosts self-confidence. ASL allows deaf children to meet deaf adults and be part of the Deaf community. "If they grow up in isolation," says Benedict, "they wouldn't know if they are capable of doing this and that."

What about cochlear implants? The logical assumption is that the access to sound that implants provide helps boost phonological awareness and that in turn boosts reading levels and academic achievement. Ken Pugh agreed this ought to be so. "Anything that provides spoken language to a child is likely to have implications for learning to read," he says. "But having said that, there are seriously good minds out there who are skeptical of that intuition. [Maybe] ultimately you can use lots of codes to support brain plasticity." For her part, Karen Emmorey, who studies the neurobiology of ASL, told me, "So much has been done looking at phonology, it's clearly key for hearing kids. It may be key for some [deaf] kids, but for some deaf kids not. . . . Something else is going on. It does seem to be that it's language that's critical."

In practice, there aren't a lot of results on reading or literacy skills beyond fourth grade in children with cochlear implants. What limited data there is hasn't been as dramatic as expected. "Not everybody ends up doing so well with reading," says Pugh, even among kids who respond well to cochlear implants.

Just as with dyslexia, the hope is that the brain will hold the answers. For example, Haskins and University of Connecticut researcher Heather Bortfeld is looking at the brains of babies and toddlers before and after they get cochlear implants and plans to follow them into elementary school and literacy. "There are a lot of open questions, but there are some principles that must span all of this," says Pugh. "The neural tissue that represents language must become reorganized in literacy." When we spoke, he had just returned from a conference in Taiwan and was embarking on a major study of literacy in four different spoken languages. "Vastly different writing systems end up with very similar patterns in the brain, because the job of the brain is to rewire and become multimodal in some sense. I think this is true whether you have input from the ear or not."

23

DEAF LIKE ME

W hen my work faltered," wrote the author Josh Swiller of a time when he was trying to write a novel while teaching at Gallaudet, "I stared at the students hanging out in the parking area. Their signing was beautiful: With their hands and arms and bodies, they carved up and fluffed out the space in front of them. They made that space breathe and filled it with energy and moved that energy around, flattening it, kneading it, passing it back and forth. . . . Watching them I thought again about what it meant to be deaf."

Reading that, I thought yes, me too. Swiller is deaf himself, but he is more like Alex than the average Gallaudet student—he was raised oral, using cued speech and hearing aids, and he got a cochlear implant as an adult. He spent a few years at Gallaudet, however, exploring other ways of being deaf.

We were about to do something similar, in a much smaller way. That was why our doorbell rang one evening shortly after dinner.

"She's here," I called and went to open the door.

A woman in her thirties with blue eyes and dark auburn hair stood on the stoop. She smiled and raised her hand in greeting. I raised mine in response and then drew my arm back to indicate the hall behind me and welcome her in.

Roni was an ASL tutor and this was to be our first lesson. Mark and the boys and our babysitter, Yvi, all congregated in the living room.

"Hi," said Jake shyly.

"Hi," said Matthew.

Then, embarrassed, they realized Roni couldn't hear them.

Again, she smiled and raised her right hand to give them all a crisp half wave, half salute and show them how to say hello.

They waved and sat down.

We were all a little unsure of ourselves. Roni was the first guest we'd ever had who communicated only in sign language.

As an ASL tutor, Roni had met many people like us and was used to our awkwardness. She took a seat in the living room and pulled out a digital tablet for writing notes and a pack of vocabulary flash cards. She waved her hand to get everyone's attention. Even to introduce ourselves, we had to learn to fingerspell the alphabet. So we started there. She wrote down the boys' names on the tablet and pointed at them quizzically to establish who was who. Then she showed them how to spell their names. When Alex's attention wandered and he looked around the room, she stamped her foot on the floor to bring his eyes back to her. She wouldn't just be teaching us to sign, she'd be teaching us how to interact with a signing deaf person. Rule number one: You have to make eye contact.

The fall from the ladder and the subsequent, though temporary, loss of what remained of Alex's hearing had goaded us into action. From the beginning, we had said that Alex would learn ASL as a second language. And we'd meant it—in a vague, well-intentioned way. Though I used a handful of signs with him in the first few months, those had fallen away once he started to talk. I was thrilled with his progress, with the cochlear implant, with the help he'd gotten at Clarke, but I regretted letting sign language lapse. I was sorry now I hadn't borrowed those baby sign language videos from my friends in Brooklyn.

Even before the additional hearing loss and in spite of my misgivings over the most recent protests at Gallaudet, I had begun to think we needed to get serious about ASL. I never wanted there to come another day when we couldn't communicate at all—in the bath or swimming pool, for instance, or if Alex's equipment malfunctioned. The fall had been a stark reminder both of how much of a difference the hearing in his left ear made and of how precarious it was. I had another reason, too. I wanted Alex to know that Deaf culture was out there.

The summer he was five, while on vacation at the beach we met a thirteen-year-old girl who had recently gotten a cochlear implant. Her mother told me her daughter could understand much of what she heard, but I could see that she still had trouble speaking. She and Alex turned their heads to show each other the processors sitting atop their ears. The girl started to sign to Alex, then stopped when she saw his blank look.

"Does he sign?" she asked.

"No," I said a little sheepishly. It was clear that I didn't, either. I couldn't even remember the sign for NO.

Even if Alex might never need ASL to communicate, he might like to know it. And he might someday feel a need to know more deaf people. At school, there were two other boys in his grade who used hearing aids, so he wasn't completely alone. All three had FM systems, which had turned out to be a microphone that accentuates the teacher's voice over classroom din, so that most of the kindergarten teachers had a transmitter slung around their necks at some point in the day. His closest friends were hearing, however. Socially, he seemed to be holding his own. The year before, one little girl had amazed me by announcing that Alex was "the funniest boy in the class." In retrospect, I realized he had earned that accolade by launching into a silly nonsensical monologue whenever he couldn't follow the conversation. Four-year-olds have a low threshold for hilarity.

"Good coping skills," remarked his teacher of the deaf, David Spritzler, who visited the classroom a few hours a week.

All the same, I wondered about his sense of self. Even I didn't know whether to call him deaf or hard of hearing, and I didn't get much help from other deaf people on that front.

"Why not say he's deaf?" one asked when I said he was hard of hearing.

But another, when I described him as deaf, said of herself, "I'm deaf; he's hard of hearing."

Either way, he was getting older, old enough to know that he was different from his father and me, different from his brothers, different from most of his peers. How would he feel about himself at six, at sixteen, at twenty-six? With whom would he feel the most affinity: hearing children, deaf children generally, deaf children with implants?

We began looking for opportunities to meet other kids like him. In New York City, there were reunions at his old school, Clarke, and events at the Center for Hearing and Communication, the organization that had worked with Caitlin Parton. At the biannual conferences put on by AG Bell, the listening and spoken language advocacy group, I could sit in on sessions about education or technology and Alex could go to a camp full of kids with implants and hearing aids.

The first morning of our first conference, when our family arrived at the hotel's restaurant for breakfast, there was someone at nearly every table with a cochlear implant.

"Wow, Alex," Jake whispered loudly. "Look!"

Alex surveyed the room. His eyes widened, then he smiled and sat up a little taller. While the boys swam and played games with the other children, I found myself gravitating to the talks in which deaf students and adults shared their experiences.

When William Stokoe first began his work classifying the structures within sign language, no one took him seriously. Surely, went the conventional wisdom, sign language was nothing more than a collection of gestures or pantomime. "Almost everyone, hearing and deaf alike, at first regarded Stokoe's notions as absurd or heretical; his books, when they came out, as worthless or nonsensical. . . . This is often the way with works of genius," wrote neurologist Oliver Sacks in *Seeing Voices*, his book on sign language and the deaf. "But within a very few years . . . the entire climate of opinion had been changed, and a revolution—a double revolution—was under way: a scientific revolution, paying attention to sign language, and its cognitive and neural substrates, as no one had ever thought to do before; and a cultural and political revolution."

Stokoe may have started that scientific revolution, but it was Ursula Bellugi who led it. She and her husband, Ed Klima, were newly married linguists in 1968, when they moved to La Jolla, California, where Klima took a job in the Department of Linguistics at the University of California, San Diego. Bellugi went to the Salk Institute for Biological Studies, where Jonas Salk, interested in expanding the scope of his research institute, had invited Bellugi and Klima to set up a small laboratory. At the time, Bellugi has said, "essentially nothing was known. . . . We had to ask new questions and invent new ways of answering them." Their scientific approach and the prestige of the Salk Institute lent the study of sign language new stature.

"They invited deaf people—children and adults, artists and poets and actors—into their lab and asked questions that even deaf people hadn't ever thought of before: What does the speed of signing mean? How are our sentences structured? What is the morphology of our words?" said Carol Padden, who worked in the lab as a graduate student. "It wasn't about helping deaf people; it was about understanding the human capacity for language. Their research legitimized deaf culture and gave us a vocabulary to discuss it, transforming the way that even deaf people talk about themselves."

For linguists, Bellugi's lab was a thrilling place and time, and over the years it attracted a long list of well-regarded researchers, deaf and hearing—so much so that many of the cognitive scientists I interviewed turned out to have worked there at some point, including Elissa Newport and Ted Supalla (Bellugi introduced them), Helen Neville, Carol Padden, Greg Hickok, and Karen Emmorey.

Bellugi and her colleagues first studied the morphological processes of ASL, establishing the richness of the internal structure of signs that provided meaning. Stokoe had made a start on this, but Bellugi and her collaborators took it further. They documented examples of morphology such as the verb LOOK-AT, which is made with the index and middle fingers pointing in a V and moving forward a few inches. Changes in the movement or orientation of the sign alter the meaning: Holding the two fingers still means STARE, circling them forward indicates that someone LOOKS-FOR-A-LONG-TIME. Tweak the sign again and you can indicate GAZE or WATCH or other variations on the theme. Bellugi also established that grammar in ASL is conveyed spatially. Pronouns such as HE, SHE, and IT are assigned specific locations. Past and future are generally indicated to the back or front of the body respectively. Noun-verb-object is conveyed in certain "directional" verbs via movement: I-HELP-YOU moves from the signer to the other person. YOU-HELP-ME moves in the opposite direction. In an important paper in 1978, Supalla and Newport showed that detailed differences in movement distinguished between a noun and a verb. Tap two fingers from one hand over the first two fingers of the other hand once and you have signed SIT. Do it twice and you have signed CHAIR.

In keeping with the biological focus of the Salk Institute and with the explosion of interest in the brain in the 1980s and 1990s, Bellugi and Klima also dug into the neurobiological foundations of sign language. The lab is credited with the groundbreaking discovery that in native speakers—deaf children of deaf parents—the brain organization of spoken and sign languages is remarkably similar. "This confirms at a neurological level that Sign *is* a language and is treated as such by the brain," wrote Sacks. Grammar and semantics are processed separately, for instance, just as they are in spoken languages. Not everything in the brain is the same, however. Helen Neville has found that an area in the right hemisphere does play a unique role in sign language, showing activity that isn't seen in spoken language. It's still unclear exactly why this area is recruited; one theory is that it enables the use of space for grammar. And there's a catch. This right-hemisphere activity is seen only in native signers, those who learn from a very early age. Like other aspects of language processing, it has a sensitive period and is not seen in anyone who learns ASL after puberty.

Research into ASL has always had a political cast, because ASL itself has such potency. Barbara Kannapell, who founded Deaf Pride in 1972, wrote: "ASL is the only thing we have that belongs to Deaf people completely." Not surprisingly, this filter colored some of the scientific work in the early days. I caught up with Carol Padden and Karen Emmorey at a conference in Boston, where both were presenting on their work, to talk about the trajectory of sign language research over the years. "Initially, it was about showing how it was the same [as spoken languages]," says Padden, who won a MacArthur "genius" grant in 2010 for her study of a Bedouin community where a high incidence of deafness led to the development of a new sign language currently in use by a fourth generation. "Now it's okay to work on how it's different from spoken language." It's also okay to embrace aspects of ASL like gesture and iconicity, the pictorial representations that do in fact exist in some ASL signs, that hewed too closely to the negative view of sign language in the past, says Emmorey. There is pantomime and gesture in sign language, she says, signing DANCE and swaying at the same time by way of example, but her work has shown that such movements use different brain areas from those employed to produce an ASL verb.

The way ASL is learned may change as well. "There's a lot of romanticism about learning through ASL," says Padden. "A lot of people say [because] they're closing the deaf schools, ASL won't have a context for people to learn the language." But she has met enough mainstreamed deaf children who've learned to sign as a second language that she's less worried on that front. "I think it's reorganizing," she says. "ASL is going to be learned in different ways. We're paying less attention to geography and more to identity."

For Emmorey, who is hearing, an interest in sign language had nothing to do with politics and everything to do with the brain. With a PhD in psycholinguistics from UCLA, she says, "I got hooked in terms of thinking about sign language as a tool to ask questions about language and the mind." When she started, she was a complete beginner at ASL, studying how sign language was organized in the brain by day and taking ASL lessons at night at a community college, the only place she could find a course. Today, at her neurobiology lab at the University of California, San Diego, all communication is in ASL. "I knew I'd arrived when I gave a lecture on cognitive neuroscience in ASL at Gallaudet . . . and people got it," she says with a laugh.

Reliable statistics on the number of people in the United States who use ASL don't really exist. The US Census asks only about use of spoken languages other than English. There are more than two million deaf people nationally, of whom between a hundred thousand and five hundred thousand are thought to communicate primarily through ASL. That equals less than a quarter of 1 percent of the national population. People have begun to throw around the statistic that ASL is the third—or fourth—most common language in the country. For that to be true, there would have to be something approaching two million users, which seems unlikely. Anecdotally, interest in ASL does seem to be growing. It is far more common as a second language in college and even high school. After Hurricane Sandy, New York City's mayor Michael Bloomberg was accompanied at every press conference by an interpreter who became a minor celebrity for her captivating signing. But people have to have more than a passing acquaintance with signing to qualify as users of the language, just as many who have some high school French would be hard-pressed to say more than *bonjour* and *merci* in Paris and can't be considered French speakers.

Still, bilingualism is the hope of the deaf community. Its leaders agree that Americans need to know English, the language of reading and writing in the United States, but they also value sign language as the "backbone" of the Deaf world. "The inherent capability of children to acquire ASL should be recognized and used to enhance their cognitive, academic, social, and emotional development," states the National Association of the Deaf. "Deaf and hard of hearing children must have the right to receive early and full exposure to ASL as a primary language, along with English."

The case for bilingualism has been helped by Ellen Bialystok, a psychologist at York University in Toronto, and the most high-profile researcher on the subject today. Her work has brought new appreciation of the potential cognitive benefits of knowing two languages. "What bilingualism does is make the brain different," Bialystok told an interviewer recently. She is careful not to say the bilingual brain is "categorically better," but she says that "most of [the] differences turn out to be advantages."

Her work has helped change old ideas. It was long thought that learning more than one language simply confused children. In 1926, one researcher suggested that using a foreign language in the home might be "one of the chief factors in producing mental retardation." As recently as a dozen years ago, my friend Sharon, whose native language is Mandarin

Chinese, was told by administrators to speak only English to her son when he started school in Houston. It is true that children who are bilingual will be a little slower to acquire both languages and furthermore, that they will have, on average, smaller vocabularies in both than a speaker of one language would be expected to have. Their grammatical proficiency will also be delayed. However, Bialystok has found that costs are offset by a gain in executive function, the set of skills we use to multitask and sustain attention and engage in higher-level thinking—some of the very skills Helen Neville was looking to build up in preschoolers and that have been shown to boost academic achievement.

In one study, Bialystok and her colleague Michelle Martin-Rhee asked young bilingual and monolingual children to sort blue circles and red squares into digital boxes—one marked with a blue square and the other with a red circle—on a computer screen. Sorting by color was relatively easy for both groups: They put blue circles into the bin marked with a blue square and red squares into the box marked with a red circle. But when they were asked to sort by shape, the bilinguals were faster to resolve confusion over the conflicting colors and put blue circles into the box with the red circle and red squares into the bin with the blue square.

When babies are regularly exposed to two languages, differences show up even in infancy, "helping explain not just how the early brain listens to language, but how listening shapes the early brain," wrote pediatrician Perri Klass in *The New York Times*. The same researchers who found that monolingual babies lose the ability to discriminate phonetic sounds from other languages before their first birthday showed that bilingual babies keep up that feat of discrimination for longer. Their world of sound is literally wider, without the early "perceptual narrowing" that babies who will grow up to speak only one language experience. Janet Werker has shown that babies with bilingual mothers can tell their moms' two languages apart but prefer both of them over other languages.

One explanation for the improvement is the practice bilinguals get switching from one language to the other. "The fact that you're constantly manipulating two languages changes some of the wiring in your brain," Bialystok said. "When somebody is bilingual, every time they use one of their languages the other one is active, it's online, ready to go. There's a potential for massive confusion and intrusions, but that doesn't happen. . . . The brain's executive control system jumps into action and takes charge of making the language you want the one you're using." Bialystok has also found that the cognitive benefits of bilingualism help ward off dementia later in life. Beyond the neurological benefits, there are other acknowledged reasons to learn more than one language, such as the practical advantages of wider communication and greater cultural literacy.

It's quite possible that some of the bias still found in oral deaf circles against sign language stems from the old way of thinking about bilingualism. It must be said, though, that it's an open question whether the specific cognitive benefits Bialystok and others have found apply to sign languages. Bialystok studies people who have two or more spoken languages. ASL travels a different avenue to reach the brain even if it's processed similarly once it gets there. "Is it really just having two languages?" asks Emmorey. "Or is it having two languages in the same modality?" Bits of Spanglish aside, a child who speaks both English and Spanish is always using his ears and mouth. He must decide whether he heard "dog" or *perro* and can say only one or the other in reply. "For two spoken languages, you have one mouth, so you've got to pick," says Emmorey. A baby who is exposed to both English and sign language doesn't have to do that. "If it's visual, they know it's ASL. If it's auditory, they know it's English. It comes presegregated for you. And it's possible to produce a sign and a word at the same time. You don't have to sit on [one language] as much." Emmorey is just beginning to explore this question, but the one study she has done so far, in collaboration with Bialystok, suggests that the cognitive changes Bialystok has previously found stem from the competition between two spoken languages rather than the existence of two language systems in the brain.

Whether ASL provides improvement in executive function—or some other as yet unidentified cognitive benefit—Emmorey argues for the cultural importance of having both languages. "I can imagine kids who get pretty far in spoken English and using their hearing, but they're still not hearing kids. They're always going to be different," she says. Many fall into sign language later in life. "[They] dive into that community because in some ways it's easy. It's: 'Oh, I don't have to struggle to hear. I can just express myself, I can just go straight and it's visual.'" She herself has felt "honored and special" when she attends a deaf cultural event such as a play or poetry performance. "It's just gorgeous. I get this [experience] because I know the language."

Perhaps the biggest problem with achieving bilingualism is the practical one of getting enough exposure and practice in two different languages. When a reporter asked Bialystok if her research meant that high school French was useful for something other than ordering a special meal in a restaurant, Bialystok said, "Sorry, no. You have to use both languages all the time. You won't get the bilingual benefit from occasional use." It's true, too, that for children who are already delayed in developing language, as most deaf and hard-of-hearing children are, there might be more reason to worry over the additional delays that can come with learning two languages at once. The wider the gap gets between hearing and deaf kids, the less likely it is ever to close entirely. When parents are bilingual, the exposure comes naturally. For everyone else, it has to be created.

I didn't know if Alex would ever be truly bilingual, but the lessons with Roni were a start. In the end, they didn't go so well, through no fault of hers. It was striking just how difficult it was for the boys, who were five, seven, and ten, to pay visual attention, to adjust to the way of interacting that was required in order to sign. It didn't help that our lessons were at seven o'clock at night and the boys were tired. I spent more time each session reining them in than learning to sign. The low point came one night when Alex persisted in hanging upside down and backward off an armchair.

"I can see her," he insisted.

And yet he was curious about the language. I could tell from the way he played with it between lessons. He decided to create his own version, which seemed to consist of opposite signs: YES was NO and so forth. After trying and failing to steer him right, I concluded that maybe experimenting with signs was a step in the right direction.

Even though we didn't get all that far that spring, there were other benefits. At the last session, after I had resolved that one big group lesson in the evening was not the way to go, Alex did all his usual clowning around and refusing to pay attention. But when it was time for Roni to leave, he gave her a powerful hug that surprised all of us.

"She's deaf like me," he announced.

24

THE COCKTAIL
PARTY PROBLEM

To my left, a boisterous group is laughing. To my right, there's another conversation under way. Behind me, too, people are talking. I can't make out the details of what they're saying, but their voices add to the din. They sound happy, as if celebrating. Dishes clatter. Music plays underneath it all.

A man standing five feet in front of me is saying something to me. "I'm sorry," I call out, raising my voice. "I can't hear you."

Here in the middle of breakfast at a busy restaurant called Lou Malnati's outside Chicago, the noise is overpowering. Until it's turned off.

I'm not actually at a restaurant; I'm sitting in a soundproof booth in the Department of Speech and Hearing Science at Arizona State University. My chair is surrounded by eight loudspeakers, each of them relaying a piece of restaurant noise. The noise really was from Lou Malnati's, but it happened some time ago. An engineer named Lawrence Revit set up an array of eight microphones in the middle of the restaurant's dining room and recorded the morning's activities. The goal was to create a real-world listening environment, but one that can be manipulated. The recordings can be played from just one speaker or from all eight or moved from speaker to speaker. The result is remarkably real—chaotic and lively, like so many restaurants where you have to lean in to hear what the person sitting across from you is saying.

The man at the door trying to talk to me is John Ayers, a jovial eighty-two-year-old Texan who has a cochlear implant in each ear. Once the recording has been switched off, he repeats what he'd said earlier.

"It's a torture chamber!" he exclaims with what I have quickly learned is a characteristic hearty laugh.

Ayers has flown from Dallas to Phoenix to willingly submit himself to this unpleasantness in the name of science. Retired from the insurance business, he is a passionate gardener (he brought seeds for the lab staff on his last visit) and an even more passionate advocate for hearing. After receiving his first implant in 2005 and the second early in 2007, he has found purpose serving as a research subject and helping to recruit other participants.

"Are you ready?" asks Sarah Cook, the graduate student who manages ASU's Cochlear Implant Lab and will run the tests today.

"Let me at it!" says Ayers.

After he bounds into the booth and takes his seat, Cook closes the two sets of doors that seal him inside. She and I sit by the computers and audiometers from which she'll run the test. For the best part of the next two hours, Ayers sits in the booth, trying to repeat sentences that come at him through the din of the restaurant playing from one or more speakers.

Hearing in noise remains the greatest unsolved problem for cochlear implants and a stark reminder that although they now provide tremendous benefit to many people, the signal they send is still exceedingly limited. "One thing that has troubled me is sometimes you hear people in the field talking about [how people have] essentially normal hearing restored, and that's just not true," says Don Eddington of

MIT. "Once one is in a fairly noisy situation, or trying to listen to a symphony, cochlear implants just aren't up to what normal hearing provides."

It wasn't until Alex lost his hearing that I properly heard the noise of the world. Harvey Fletcher of Bell Labs described noise as sounds to which no definite pitch can be assigned, and as everything other than speech and music. Elsewhere, I've seen it defined as unwanted sound. The low hum of airplane cabins or car engines, sneakers squeaking and balls bouncing in a gym, air conditioners and televisions, electronic toys, a radio playing, Jake and Matthew talking at once, or tap water running in the kitchen. All of it is noise and all of it makes things considerably harder for Alex. Hearing aids aren't selective in what they amplify. Cochlear implants can't pick and choose what sounds to process. So noise doesn't just make it harder to understand what someone is saying; ironically, it can also be uncomfortably loud for a person with assistive devices. Some parents of children with implants or hearing aids stop playing music completely at home in an effort to control noise levels. Many people with hearing loss avoid parties or restaurants. We haven't gone quite that far, but I was continually walking into familiar settings and hearing them anew.

To better assess how people with hearing loss function in the real world, audiologists routinely test them "in noise" in the sound booth. The first time Lisa Goldin did that to Alex was the only time he put his head down on the table and refused to cooperate. She was playing something called "multitalker babble," which sounded like simultaneous translation at the United Nations. Even for me, it was hard to hear the words Alex was supposed to pick out and repeat. Lisa wasn't trying to be cruel. An elementary school classroom during lunchtime or small group work can sound as cacophonous as the United Nations. Even then, hearing children are learning from one another incidentally. Like so much else in life, practice would make it easier for Alex to do the same—though he'll never get as much of this kind of conversation as the others—and the test would allow Lisa to see if there was any need to adjust his sound processing programs to help.

Hearing in noise is such a big problem—and such an intriguing research question—that it has triggered a subspecialty in acoustic science known as the "cocktail party problem." Researchers are asking, How does one manage to stand in a crowd and not only pick out but also understand the voice of the person with whom you're making small talk amid all the other chatter of the average gathering? Deaf and hard-of-hearing people have their own everyday variation: the dinner table problem. Except that unlike hearing people at a party, deaf people can't pick out much of anything. Even a mealtime conversation with just our family of five can be hard for Alex to follow, and a restaurant is usually impossible. His solution at a noisy table is to sit in my lap so I can talk into his ear, or he gives up and plays with my phone—and I let him.

"If we understand better how the brain does it with normal hearing, we'll be in a better position to transmit that information via cochlear implants or maybe with hearing aids," says Andrew Oxenham, the auditory neuroscientist from the University of Minnesota. Intriguingly, understanding the cocktail party problem may not only help people with hearing loss but could also be applied to automatic speech recognition technology, too. "We have systems that are getting better and better at recognizing speech, but they tend to fail in more complicated acoustic environments," says Oxenham. "If someone else is talking in the background or if a door slams, most of these programs have no way of telling what's speech and what's a door slamming."

The basic question is how we separate what we want to listen to from everything else that's going on. The answer is that we use a series of cues that scientists think of as a chain. First, we listen for the onset of new sounds. "Things that start at the same time and often stop at the same time tend to come from the same source. The brain has to stream those segments together," says Oxenham. To follow the segments over time, we use pitch. "My voice will go up and down in pitch, but it will still take a fairly smooth and slow contour, so that you typically don't get sounds that drastically alter pitch from one moment to the next," says Oxenham. "The brain uses that information to figure out, well, if something's not varying much in pitch, it probably all belongs to the same source." Finally, it helps to know where the sound is coming from. "If one thing is coming from the left and one thing is coming from the right, we can use that information to follow one source and ignore another."

The ability to tell where a sound is coming from is known as spatial localization. It's a skill that requires two ears. Anyone who has played Marco Polo in the pool as a child will remember that people with normal hearing are not all equally good at this, but it's almost impossible for people with hearing loss. This became obvious as soon as Alex was big enough to walk around the house by himself.

"Mom, Mom, where are you?" he would call from the hall.

"I'm here."

"Where?"

"Here."

Looking down through the stairwell, I could see him in the hall one floor and perhaps fifteen feet away, looking everywhere but at me.

"I'm here" wouldn't suffice. He couldn't even tell if I was upstairs or downstairs. I began to give the domestic version of latitude and longitude: "In the bathroom on the second floor." Or "By the closet in Jake's room."

To find a sound, those with normal hearing compare the information arriving at each ear in two ways: timing and intensity. If I am standing directly in front of Alex, his voice reaches both of my ears simultaneously. But if he runs off to my right to pet the dog, his voice will reach my right ear first, if only by a millionth of a second. The farther he moves to my right, the larger the difference in time. There can also be a difference in the sound pressure level or intensity as sounds reach each ear. If a sound is off to one side, the head casts a shadow and reduces the intensity of the sound pressure level on the side away from the source.

Time differences work well for low-frequency waves. Because high-frequency waves are smaller and closer together, they can more easily be located with intensity differences. At 1,000 Hz, the sound level is about eight decibels louder in the ear nearer the source, but at 10,000 Hz it could be as much as thirty decibels louder. At high frequencies, we can also use our pinna (the outermost part of the ear) to figure out if a sound is in front of us or behind. Having two ears, then, helps with the computations our brain is constantly performing on the information it is taking in. We can make use of the inherent redundancies to compare and contrast information from both ears.

Hearing well in noise requires not just two ears but also a level of acoustic information that isn't being transmitted in today's implant. A waveform carries information both in big-picture outline and in fine-grained detail. Over the past ten years, sound scientists have been intensely interested in the difference, which comes down to timing. To represent the big picture, they imagine lines running along the top and bottom of a particular sound wave, with the peaks and troughs of each swell bumping against them. The resulting outline is known as the envelope of the signal, a broad sketch of its character and outer limits that captures the slowly varying overall amplitude of the sound. What Blake Wilson and Charlie Finley figured out when they created their breakthrough speech processing program, CIS, was how to send the envelope of a sound as instructions to a cochlear implant.

The rest of the information carried by the waveform is in the fine-grained detail found inside the envelope. This "temporal fine structure" carries richness and depth. If the envelope is the equivalent of a line drawing of, for example, a bridge over a stream, fine structure is Monet's painting of his Japanese garden at Giverny, full of color and lush beauty. The technical difference between the two is that the sound signal of the envelope changes more slowly over time, by the order of several hundred hertz per second, whereas "fine structure is the very rapidly varying sound pressure, the phase of the signal," says Oxenham. In normal hearing, the fine structure can vary more than a thousand times a second, and the hair cells can follow along.

An implant isn't up to that task. So far, researchers have been stymied by the limits of electrical stimulation—or more precisely by its excesses. When multiple electrodes stimulate the cochlea, in an environment filled with conductive fluid, the current each one sends spreads out beyond the intended targets. Hugh McDermott, one of the Melbourne researchers, uses an apt analogy to capture the problem. He describes the twenty-two electrodes in the Australian cochlear implant as twenty-two cans of spray paint, the neurons you're trying to stimulate as a blank canvas, and the paint itself as the electrical current running between the two. "Your job in creating a sound is to paint something on that canvas," says McDermott. "Now the problem is, you can turn on any of those cans of paint anytime you like, but as the paint comes out it spreads out. It has to cross a couple meters' distance and then it hits the canvas. Instead of getting a nice fine line, you get a big amorphous blob. To make a picture of some kind, you won't get any detail. It's like a cartoon rather than a proper painting." In normal hearing, by contrast, the signals sent by hair cells, while also electrical, are as controlled and precise as the narrowest of paintbrushes.

So while it seems logical that more electrodes lead to better hearing, the truth is that because of this problem of current spread, some of the electrodes cancel one another out. René Gifford, of Vanderbilt University, is working on a three-way imaging process that allows clinicians to determine—or really improve the odds on guessing—which electrodes overlap most significantly, and then simply turn some off. "Turning off electrodes is the newest, hottest thing," says Michael Dorman of Arizona State University, who shared Gifford's results with me. Gifford is a former member of Dorman's laboratory, so he's rooting her on. Half of those she tested benefited from this strategy. Other researchers are working on other ideas to solve the current-spread problem. Thus far, the best implant manufacturers have been able to do is offer settings that allow a user to reduce noise if the situation requires it. "It's more tolerable to go into noisy environments," says Dorman. "They may not understand anything any better, but at least they don't have to leave because they're being assaulted."

In a handful of labs like Dorman's, where Ayers and Cook are hard at work, the cocktail party problem meets the cochlear implant. "You need two ears of some kind to solve the cocktail party problem," says Dorman, a scientist who, like Poeppel, enjoys talking through multiple dimensions of his work. For a long time, however, no one with cochlear implants had more than one. The reasons were several: a desire to save one ear for later technology; uncertainty about how to program two implants together; and—probably most significant—cost and an unwillingness on the part of insurance companies to pay for a second implant. As I knew from my experience with Alex, for a long time it was also uncommon to use an implant and hearing aid together. Within the past decade, and especially the past five years, that has changed dramatically. If a family opts for a cochlear implant for a profoundly deaf child, it is now considered optimal to give that child two implants simultaneously at twelve months of age—or earlier. In addition, as candidacy requirements widen, there is a rapidly growing group of implant users with considerable residual hearing in the unimplanted ear. Some even use an implant when they have normal hearing in the other ear.

Dorman's goal has been to put as many people as possible with either two cochlear implants ("bilaterals") or with an implant and a hearing aid ("bimodals") through the torture chamber of the eight loudspeaker array to look for patterns in their responses, both to determine if two really are better than one and, if so, to better understand how and why.

For John Ayers, Cook doesn't play the restaurant noise as loud as it would be in real life. With the click of a computer mouse, she can adjust the signal-to-noise ratio—the relative intensity of the thing you are trying to hear (the signal) versus all the distracting din in the background (the noise). She makes the noise ten decibels quieter than the talker, even though the difference would probably be only two decibels in a truly noisy restaurant. She needs first to establish a level at which Ayers will be able to have some success but not too much, so as to allow room for improvement. Noise that's so loud he can't make out a word or so quiet he gets everything from the start doesn't tell the researchers much. Eventually, Cook settles on a level that is six decibels quieter than the signal. Ayers repeats the test with one implant, then the other, then both together—each time trying different noise conditions, with the noise coming from just one loudspeaker or from all of them. From the computers where I sit, it's hard to see him through the observation window, so he's a disembodied voice saying things like, "He was letting Joe go," when it should have been, "He went sledding down the hill."

The sentences Cook asks Ayers to repeat were created in this very lab in an effort to improve testing by providing multiple sentence lists of equivalent difficulty. Known as the AzBio sentences, there are one thousand in all recorded by Dorman and three other people from the lab. They're widely used. That meant, back home in New York City, I could still hear Dorman's deep, sonorous voice speaking to me when I observed a test session in Mario Svirsky's laboratory. To relieve the tedium of the sound booth, Dorman and colleagues intentionally made some of the sentences amusing.

"Stay positive and it will all be over." Ayers got that one.

"You deserve a break today." Ayers heard it as: "You decided to fight today."

"The pet monkey wore a diaper." A pause and then Ayers says, incredulously: "Put the monkey in a diaper?"

Cook scores the sentences based on how many words Ayers gets right out of the total. With only one implant, Ayers scored between 30 and 50 percent correct. With both implants together, he scored as high as 80 percent.

Dorman and Cook use the same loudspeaker array to test the ability of cochlear implant users to localize sound, which is restored by two implants but only to a degree, as implants can work with intensity cues but not timing. Hearing aids, on the other hand, can handle timing cues, since the residual hearing they amplify is usually in the low frequencies. The average hearing person can find the source of a sound to within seven degrees of error. Bilateral implant patients can do it to about twenty degrees. "In the real world, that's fine," says Dorman. It works because the bilateral patients have been given the gift of a head shadow effect. "If you have two implants, you'll always have one ear where the noise is being attenuated by the head," says Dorman. He sees patients improve by 30 to 50 percent.

With both a hearing aid and a cochlear implant, Alex uses two ears, too, so it seemed he ought to have had an easier time localizing sound than he did. During my visit to Arizona, I finally understood why localizing was still so hard for him. Bimodal patients—those with an implant and a hearing aid—do better than people with just one usable ear, who can't localize at all, but the tricks that the brain uses to analyze sound coming into two different ears require something bimodal patients don't have: two of the same kind of ears. "Either will do," says Dorman. "For this job of localizing, you need two ears with either good temporal cues or good intensity cues." A hearing aid gives you the first, an implant gives you the second, but the listener with one of each is comparing apples to oranges.

The work with bilateral and bimodal patients is a sign of the times. The basic technology of implants hasn't actually changed much in twenty years, since the invention of CIS processing. Absent further improvements in the processing program or solutions to the problem of spreading electrical current, the biggest developments today have less to do with how implants work and more with who gets them, how many, and when. Just because the breakthroughs are less dramatic these days, says Dorman, that doesn't mean they don't matter. He has faith in the possibilities of science and says, "You have to believe that if we can keep adding up the little gains, we get someplace." One of the projects he is most excited about is a new method that uses modulation discrimination to determine if someone like Alex would do better with a hearing aid or a second implant. "It allows you to assess the ability of the remaining hearing to resolve the speech signal. So far, it's more useful than the audiogram." The project is still in development so won't be in clinical use for several years, but the day they realized how well the strategy worked was a happy one. "You keep playing twenty questions with Mother Nature and you usually lose," says Dorman. "Every once in a while, you get a little piece of the answer, steal the secret. That's a good day."

25

BEETHOVEN'S NIGHTMARE

Alex waved with delight, thrilled to see me in the middle of a school day. Head tilted, lips pressed together, big brown eyes bright, he wore his trademark expression, equal parts silly and shy. His body wiggled with excitement. I waved back, trying to look equally happy. But I was nervous. Alex and the other kindergartners at Berkeley Carroll were going to demonstrate to their parents what they were doing in music. Three kindergarten classes had joined forces, so there were nearly sixty children on the floor and at least as many parents filling the bleachers of the gym, which doubled as a performance space.

It had been almost exactly three years since Alex's implant surgery. Now he was one of this group of happy children about to show their parents what they knew about pitch, rhythm, tempo, and so on. Implants, however, are designed to help users make sense of speech. Depending on your perspective, music is either an afterthought or the last frontier. Or was. Some of the same ideas that could improve hearing in noise might also make it possible for implant users to have music in their lives. I was thrilled to know that people were out there working on this, but they couldn't help Alex get through kindergarten.

Music appreciation and an understanding of its basic elements were among the many pieces of knowledge he and the other children were expected to acquire. I feared—even assumed—music was one area where his hearing loss made the playing field too uneven.

Music is much more difficult than speech for the implant's processor to accurately translate for the brain. As a result, many implant recipients don't enjoy listening to music. In her account of receiving her own implant, *Wired for Sound*, Beverly Biderman noted that for some recipients, music sounded like marbles rolling around in a dryer. After she was implanted, Biderman was determined to enjoy music and worked hard at it. (Training does help, studies show.) For every twenty recordings Biderman took out of the library to try, eighteen or nineteen sounded "awful," but one or two were beautiful and repaid her effort.

Speech and music do consist of the same basic elements unfolding over time to convey a message. Words and sentences can be short or long, spaced close together or with big gaps in between—in music we call that rhythm. The sound waves of spoken consonants and vowels have different frequencies and so do musical notes—that's pitch. Both spoken and musical sounds have what is known as "tonal color," something of a catchall category to describe what's left after rhythm and pitch—timbre, the quality that allows us to recognize a voice or to distinguish between a trumpet and a clarinet.

But music is far more acoustically complex than speech, and its message is abstract. "There's a big difference between what music expects of hearing and what speech requires," acoustic scientist Charles Limb of Johns Hopkins University explains to me. "Speech is redundant. It's all within a certain frequency range." So speech doesn't require as much information to make sense of it. The words themselves are a handy clue, as is the context in which they are used. Musical sounds have a lot more going on within them, and if there are no lyrics to serve as guideposts, it gets even harder. Classical music is generally much harder to follow than pop, for instance.

In one of his papers, Limb compared "Happy Birthday to You" when spoken, sung, or played on the piano. Represented in waveform, the spoken words are distinct, narrow bands. When sung or played on the piano, those same bands begin to spread out and "smear" like a squirt of ketchup after it's been mushed into a hot dog roll. For all of us, then, the sound waves of music are smeared already. A cochlear implant smears the sound further and the result can be a muddle.

There are a handful of people, however, who defy expectations. In Melbourne, Australia, there's a young woman who lost her hearing suddenly around the age of thirty. Before that, she played the piano very regularly. Within eighteen months of going deaf, she got a cochlear implant and now has two. "She still continues to play the piano a few hours a day," says Peter Blamey, who has been working with cochlear implant recipients since he joined Graeme Clark's team in 1979. Now a deputy director at the Bionics Institute, a research organization founded (as the Bionic Ear Institute) by Clark, Blamey says of this woman, "She has pitch perception that's as good as mine. She can do things like rating consonance and dissonance of chords and notes played in succession and do things that are generally not accepted as being possible with a cochlear implant." Blamey and his colleagues are still studying her, but they are guessing that in her case training has been the important factor. "The difference seems to be that six hours a day that she spends playing the piano, and maybe the learning that took place before she went deaf. They're both central things," he says. "We're looking for things that are going to be more generally applicable for improving music perception for people even if they don't have six hours a day to devote to it."

Researchers at the Bionics Institute are also using the same cues that Andrew Oxenham described to me as useful for hearing in noise to see if they can tease out ways to use them to make listening to music not just more accurate but more enjoyable for people with implants. In their research, "we just have simple musical melodies that repeat over and over that the brain can learn easily," explains research scientist Hamish Innes-Brown. "Then we vary those notes in loudness, pitch, location, etc. We try to get people to detect that modification. We want to change the signal as little as possible—get the most streaming bang for your perceptual buck." They also recently commissioned composers to create music specifically for cochlear implant users. The musicians spent nine months or so visiting the institute and learning about implants. One named Natasha Anderson impressed Innes-Brown by asking for the center frequencies of all the implants. "She's actually tailored this sound to fit all the exact frequencies."

By the day of the open kindergarten music class, I had spent a lot of time considering what words Alex could understand and say, but I hadn't thought deeply about music and how he experienced it, except that he loved it. Maybe he liked the sound of marbles in the dryer—the little rascal wasn't above putting marbles in the dryer. But it was more likely that that's not what it sounded like to him. Music and dancing were such favorites that at Clarke his end-of-year gift had been a book called *Song and Dance Man*. He has a hard time keeping up with song lyrics, but his favorite family activity for years was a "dance party," which entailed putting on music and having all of us boogie around the living room.

"Does he respond to music?" was one of the questions the doctor asked when we were trying to figure out why he couldn't talk.

"Yes," we had to say, "with gusto."

When I began to look into research on music and cochlear implants, the mystery was explained. There was one group that consistently reported greater enjoyment of music: those who still had some hearing in the non-implanted ear and used a hearing aid. Furthermore, even profoundly deaf people enjoy the vibration and beat of music. The all-deaf rock band Beethoven's Nightmare was featured in the PBS documentary *Through Deaf Eyes*. The group's drummer, Bob Hiltermann, said he depended on both vibrations and his hearing and that the band played "really, really loud" so that the musicians could hear themselves. He joked about attending a rock concert: "It was really too loud for the regular hearing person," he said. "They're going to become deaf themselves. But we already are, so it's perfect." In a lecture on how to listen to music with your whole body, deaf percussionist Dame Evelyn Glennie remembered what she said to her first teacher when he asked how she would hear the music: "I hear it through my hands, through my arms, my cheekbones, my scalp, my tummy, my chest, my legs, and so on." Glennie even plays without shoes, the better to hear the music with her feet.

That, I suspect, was some of what Alex experienced when he danced around the living room. He didn't hear everything that we heard, but he loved what he did hear. He listened with his whole body. Music brought him joy, just as it had millions of other people over the centuries. But dancing around the living room is not the same thing as performing music in class. At school, there was an audience. The children would have to demonstrate knowledge and skill. Would performing music in this way kill the joy for Alex? Was I setting him up to fail? At a demoralizing dance performance, he was once the unfortunate child whose partner didn't show up, and though he gamely went through the routine with the rest of the class, he was behind from the start and never caught up.

I read through the music teacher's three-page description of what the children were going to do like a doctor searching a medical history for signs of trouble. Trouble showed itself quickly in the section about the importance of "inner hearing":

> Inner hearing in music is an essential tool needed to sing in tune, read music, create musical compositions and to improvise melodies. The Kindergartners are able to tap beats

while singing melodies in their heads. Honing the inner voice is not only essential for the professional musician but for the lifelong lover of music.

Was my child able to do that?

And what about this part about how "once children are able to clap the rhythm of the words to the song, transferring the rhythm to a percussion instrument is a given."

Was it a given for Alex?

Maybe. Hands down, researchers have found, rhythm is easiest for implant recipients to perceive. They do nearly as well in listening tests as those with normal hearing. That's because the information that has to be processed occurs over seconds, not milliseconds, and it doesn't require the fine-tuning that other musical elements do. With each beat, an electrode will fire. Discerning the exact frequency doesn't matter. This is why percussive instruments such as drums and piano are easiest for implant recipients to appreciate. (And probably why, when Beverly Biderman embarked on her own musical study experiment, the first music that sounded good to her was Glenn Gould's recording of *The Goldberg Variations* on the piano.)

Pitch and timbre are much more difficult. Is one pitch higher or lower than another? Some implant users can answer that question, but some can't identify even an octave change. Pitch is essential for following a melody, but not nearly so important in understanding a sentence. It will tell me whether I'm listening to a man, woman, or child, but I don't absolutely need to know who is speaking to know that he or she said "Excuse me" or "May I have some water?" In music, however, the only way I can differentiate between "Twinkle, Twinkle Little Star" and "Mary Had a Little Lamb" is by following the changes in pitch. Pitch is essential to melody and melody is essential to appreciating music.

Like localization, pitch is an area where Alex's two modes of hearing might not always be helpful. His low-frequency hearing in the left ear probably conveys some pitches fairly accurately, but his implant may tell him something completely different about the same note—the frequency shift that Mario Svirsky had explained to me. If you compare the sound frequency spectrum to a rainbow, an implant recipient can see the ROYGBIV colors—red, orange, yellow, green, blue, indigo, and violet—but he or she cannot see any of the colors in between, Charles Limb explained. Furthermore, the information they get may not be accurate—the equivalent of seeing orange as red.

In studies of timbre, less than half of implant recipients were able to correctly identify musical instruments. Non-musicians make mistakes, too, but their mistakes are less frequent and tend to be within the same instrumental family, such as confusing an oboe with a clarinet. Implant recipients might mistake a flute for a trumpet, a violin for an organ.

At least I knew that Alex should have no problem with the drums in the music class. They were arrayed in rows on the floor of the gym. The xylophones were lined up behind them. Looking around the gym, a space I had been in many times over several years, I noticed how small movements in the bleachers rang out loudly and how the noise of the children as they settled into their spots on the floor was like a low roar. Static or feedback in the public address system was suddenly not just annoying but worrisome. (The year before, I'd been at a memorial service where the microphone didn't work well and two family friends with hearing aids—men in their seventies—told me they couldn't understand a word that was said.)

After some discussion on the importance of being quiet in music class—something for which I had new appreciation—the group started in on the traditional standard "Engine, Engine Number Nine." With sixty-some kindergartners chanting at different tempos—fast and slow—the gym qualified as an "in noise" condition. Just as Usha Goswami had recommended, the children were playing with rhythm and rhyme and having a great time doing it. Alex seemed as happy as anyone.

Next, the kids took turns at easels where sheets of pictures—a frog, for instance, repeated in rows and columns—helped them tap the beat. The teacher had devised a game in which children tried to tap with the chant and end on the last picture at exactly the last beat of the song. I wondered if Alex could distinguish the separate words of the song to help him distinguish the beats. In the same way that other children think "LMNO" is one letter when they are first learning the alphabet, he often has trouble separating sounds. In general, though, tapping a beat like this should be quite possible for Alex. And it was.

After a few variations of this and some drumming and xylophone playing, we were nearly at the end of the class. I was beginning to think I shouldn't have worried so much. For the last exercise, the teacher mentioned the inner hearing he had written about in the handout. He was going to sing a melody. The children had to listen and then play it back on a two-tone wood block, improvising as they decided which tone went with which beat. Hands shot up around the circle of children to volunteer, Alex's among them. At least he's enthusiastic, I thought. The teacher picked a little girl from the right side of the circle, and then, from the left side, he picked Alex.

Stunned, I sat up sharply. What could he be thinking? Inner hearing, he said. Inner *hearing*! This teacher had just chosen Alex to demonstrate an activity that was all about hearing, in front of more than one hundred people. I knew that this particular teacher had taken a special interest in Alex. As a musician, he was naturally interested in sound. Did he know something I didn't know? Had he practiced with Alex? I didn't think so.

The little girl went first. The teacher sang:

Bah, bah, ba-di-di, bah,
Bah, bah, ba-di, ba-di, bah.

The girl picked up her wood block and played it back perfectly. After a round of applause, she returned to her place in the circle.

Now the only child standing in the middle was Alex. The teacher sang a slight variation:

Bah-di, bah-di, ba-bah,
Bah-di, bah-di, ba-ba-bah.

Alex picked up his wood block and played it back a little more tentatively than his classmate:

Bah-di, bah-di, ba-bah.

Slight hesitation:

Bah-di, bah-di, ba-ba-bah.

The teacher started applauding just before Alex reached the end—the only sign that he might have been nervous about putting Alex on the spot. Alex gave a shy smile and trotted back to his seat.

He did it! The lowly wood block was as spectacular to me in that moment as a Steinway grand set up in Carnegie Hall. Around me, everyone was clapping and murmuring approval. "He did great," the mother sitting next to me said. But I don't have the sense that anyone else quite understood the significance. It was bigger than Alex somehow. It called to mind the decades of work and dozens of people who had made it possible for my child to have this experience.

Not far from the foot of the bleacher steps, Alex assembled with his class, each kid waiting to hug a mother or father before they returned to their room. I was trying to keep myself collected. When I reached the gym floor, he ran into my arms. "I am so proud of you," I said into his ear and held him for a few seconds longer than normal. Then we said good-bye.

I made for the door, trying to avoid conversation and weaving through the other parents still milling around. Then I saw a woman coming against traffic, heading straight toward me instead of the door. She caught my emotional eye. It was a mother whose son had been identified with moderate hearing loss the year before. He now had hearing aids, and I had helped guide her through the early steps in the process—both bureaucratic and human. She stopped in front of me.

"That was pretty amazing," she said with a grin. She put her arms around me. "Yes," I mumbled into her shoulder, "it was."

26

WALK BESIDE ME

The Marketplace cafe at Gallaudet University sits at the bottom of a two-story atrium. The large center staircase and the railings above make it possible for students and teachers to communicate across long distances. And they do. From the stair, one young woman is waving, using her entire wingspan, to get the attention of a friend on the floor below. Some students are flirting, a girl coyly showing a boy something on her cell phone and laughing. A young man at another table is telling his friends a story and everyone is rapt. A *Looney Tunes* cartoon is playing on the television mounted to a pillar above me. As I eat my lunch and check my e-mail, I find the soundtrack annoying and, for a moment, I wonder why they don't shut it off. Then I remember that I am one of very few people in the cafe who can hear it.

Since it's a lovely September afternoon and I have time, I decide to find a bench in the sun on the green running through the middle of campus—the whole place is known as Kendall Green after the property's original owner, Amos Kendall. On the way, I pass the community bulletin boards. The postings are exactly the same as at every other college campus (a basketball fund-raiser, coming-out stories, a bike-share program, a religious group advertising a talk entitled "What Am I Doing Here?") and utterly unique (a Deaf history lecture series featuring Deaf heroes of World War II and the lives of Deaf photographers, a chance to be in Deaf America's Got Talent, a lecture on reducing split visual attention in the classroom, an information session on studying abroad in Italian Sign Language). Among the campus happenings are advertisements that share the theme of accessibility, like a local hairstylist who has learned ASL and a mechanic whose business is called Deafwrench's Garage.

Outside, young people pass me in Go Bison sweatshirts—the mascot was adopted in the 1940s for its power and fleetness. A couple holding hands lets go briefly to communicate and then grabs hold again. A fraternity trying to drum up members is frying bratwurst outside the student center. The student body is notably diverse, with more students in wheelchairs than you would see elsewhere and a few who are both deaf and blind. Cell phones are in everyone's hands, but no one holds one to an ear—they're for texting and e-mail only, exactly as hearing teenagers I know use them. Here, too, I'm aware of how much sound there is. A lawn mower thrums nearby. Sirens blare on Florida Avenue. (Earlier, and alarmingly, I'd seen a student nearly get hit by a fire truck.) There's an occasional bark of deaf laughter from the bratwurst table. Perhaps the noise is heightened by the quiet that otherwise surrounds me.

This was my second visit to Gallaudet. A few months earlier, in the summer, I had spent two weeks on campus in an intensive ASL course. My journal entry the day I arrived says simply: "I am here." Since those early nights when I was trolling the Internet for information on hearing loss, the place had loomed large as the center of Deaf culture, with what I presumed would be a correspondingly large number of cochlear implant haters. By the time I got there in 2012, I no longer imagined I would be turned back at the front gates, but just the year before a survey had shown that only one-third of the student body believed hearing parents should be permitted to choose implants for their deaf children.

The current conversation is much more subtle. It's less about cochlear implants and more about the value and endurance of ASL and Deaf culture and of "a visual way of living." In 2000, the National Association of the Deaf revised its position on cochlear implants to declare them "a technology that represents a tool to be used in some forms of communication, and not a cure for deafness." The number of Gallaudet students with cochlear implants stands at 10 percent of undergraduates and 7 percent overall. But implants still stir strong emotion. As a hearing person paying a few brief visits, I could only hope to scratch the surface of the truth of Gallaudet today. I would try at least to do that.

A place of deep group connection, Gallaudet is also these days a place of internal conflict, full of the sort of hushed, heated discussion kept strictly within the family. In recovery from the protests of 2006 and the threat to its accreditation that followed, the school is trying to determine the way forward. In a world where only 5 percent of deaf and hard-of-hearing children are born into Deaf culture by virtue of having deaf parents, demographics are not in its favor. It's generally accepted that half of cochlear implant recipients are children. In 2009, it was estimated that of the nine to ten American children born deaf every day, 40 percent get at least one and quite possibly two cochlear implants before they turn three. In Australia, 80 percent of deaf children have implants. Current statistics are hard to find, but a statewide early-intervention organization in North Carolina called Beginnings, which offers remarkably evenhanded information about communication choices, has found that the percentage of families it serves who choose listening and speaking has hovered around 90 percent over the past few years, up from 69 percent in 2001. Correspondingly, the number choosing total communication has dropped from 21 percent to 8 percent, and ASL alone is less than that. As for adults, those who lose their hearing later in life are far less likely to ever become part of Deaf culture. I. King Jordan, the Gallaudet professor who became the university's first deaf president, for instance, lost his hearing after a motorcycle accident at the age of nineteen. Someone like that would probably get a cochlear implant today.

Josh Swiller, who wrote a searing and beautiful book, *The Unheard*, on his experience growing up with hearing loss and serving in the Peace Corps in Africa, found Gallaudet an "uncomfortable" place when he first arrived as an instructor. Mostly appreciative of his hearing aids and then cochlear implant, he didn't learn to sign until he was an adult. His view of the world was different from many of his students'. In one of the first courses he taught at the university, examining minority cultures at times of crisis through history, he challenged the class by telling them: "The world [is] changing and changing away from you. So what will you do? What will you offer? Why should the world care?" Painful though it is, there is truth there. Deaf culture is a small group really, and deaf of deaf—the deaf children of deaf parents—though most interesting to neuroscientists and social scientists, are least representative of deaf people as a whole. At Gallaudet, however, they are the elite. They set the tone.

In a magazine article recounting his first year on campus and specifically his role as a mentor to a brilliant but erratic deaf-of-deaf aspiring writer, Swiller described a place that is roiled in debate over whether survival requires emphasis on communal connection or on preparing students to find jobs. His own view is that economics trump emotion, yet he found Gallaudet's ethos unexpectedly appealing. Despite the many achievements of deaf people as doctors, lawyers, professional athletes, Academy Award winners, he wrote, "the world hears and expects you to hear. . . . At some point, if you're deaf, every accomplishment fades away and you're sitting in the corner, lost. What I saw was that inside the gates of Gallaudet, everyone's been in that corner. Some have raged against it, some have ignored it, some have found spiritual riches in the surrender to limitation, some have felt cheated by it. And from that shared disconnect there has stemmed a connection that is the essence of Gallaudet. It's a gorgeous thing—many people with no hearing loss at all come to Gallaudet to be part of it."

Few of the people who made up my summer ASL class had any hearing loss. Most were college students studying ASL as a second language. One wanted to be a disability rights attorney, another a speech pathologist, and two men were in seminary studying to be priests. I was the only one there because of my child.

Our teacher, Janis Cole, was a warm and funny woman in her fifties, tall and athletic. I liked her from the beginning. Early on, she held a hand to her mouth and turned a pretend key—TURN OFF YOUR VOICE. It wasn't just that we were in an immersion program. There is a code of behavior on campus that requires that all communication be signed—by someone. If you can't do it yourself, you make sure someone is interpreting for you. That was difficult for our class of beginners to pull off and made for limited conversation at lunch (until, to be frank, people gave up). I felt frustration at not being able to express myself fully, and also chagrin for flubbing it so badly in the first day or two, before I really understood the etiquette.

By the last day of class, with enough vocabulary and a smattering of fingerspelling under our belts, Janis thought we were ready for some entertainment. ASL requires expressiveness—some of its grammar is read on the face—but earlier in life, Janis had been an actor with the National Theatre of the Deaf and appeared in *Children of a Lesser God* on Broadway, so her signing was particularly theatrical.

Standing in the middle of the circle of desks, she rubbed one fist on top of the other twice with her forefingers half-crooked. JOKE. "A lumberjack goes into the woods," she signed. "He sees a tree, chops at its trunk, yells 'Timber!' and the tree falls down. Then he chops another tree, yells 'Timber!' again, and the second tree falls down." Janis placed her right arm upright on top of her left hand and wiggled her right fingers. TREE. Then her right hand fell to her left elbow as the tree fell. "The lumberjack tries to cut down a third tree, but this time the tree won't fall. 'Timber!' he cries. 'Timber!' Nothing. So he goes to find a doctor. 'Doctor, I chopped down a tree but it won't fall. What's wrong with it?' The doctor joins the lumberjack by the tree and examines it. Then he looks up at the tree and fingerspells T-I-M-B-E-R. The tree falls. The doctor turns to the lumberjack and says, 'The tree is deaf.'" We all laughed and I realized it was the first deaf joke I'd experienced as it was meant to be received—visually.

In my class, I was a parent. But on campus, I was also a journalist. Before arriving, I had written to several faculty and administrators asking to meet with them while I was there. For a variety of reasons I could mostly only guess at, many said no, including unfortunately the scientists heading up the Visual Language Visual Learning laboratory, known as VL2, who work with Rochester's Peter Hauser and whose research on how the brain processes visual language was of great interest to me.

Stephen Weiner was different. Although he has held jobs elsewhere, Weiner has spent more of his adult life than not at Gallaudet since he first arrived in the 1970s. He has been a student, residential advisor, guidance counselor, professor, and dean. Today, he is the university's provost, the chief academic officer. His brother, Fred Weiner, also works for Gallaudet and was one of the alumni leaders of the 1988 Deaf President Now protest. From his first e-mail, Steve Weiner was friendly and approachable, eager to fill me in on all that is going on at Gallaudet, such as VL2 and the push for Deaf Space, architecture designed with natural light and good sight lines in mind and exemplified by the new Sorenson Language and Communication Center. A self-described "Jewish kid from Flatbush," he was happy to talk about his own experiences, too. He closed that first message: "Welcome to Kendall Green."

In person, greeting me with a hearty handshake and a breathy "I'm Steve," he ushers me into his office, with its view of the United States Capitol. We settle around the coffee table with an interpreter by my side, and we start at the beginning. It is clear that Weiner has seen the good and the bad of deaf education. His parents were the children of Orthodox Jewish immigrants. His grandmother became a powerful local politician in Brooklyn who was strenuously opposed to having her son, Weiner's father, use sign language. Instead, the boy moved through various New York schools, such as Lexington and P.S. 47, a public school for the deaf, missing much of what was said. "My father's education ended at third or fourth grade," says Weiner. "He [also] went through his Bar Mitzvah knowing bubkes about what was going on." Like so many deaf children, Weiner's father learned to sign from deaf friends. "My father is not educated, but I thought he was one of the smartest men I knew. He can sign, but he didn't have the vocabulary to express things. He told me later how frustrating that was."

When Stephen was born, his grandfather, having sat back and watched his son struggle, decided to do things differently this time. "My grandmother was a politician, my grandfather was pragmatic," he says. "He did not want to make the same mistakes with me that he made with my father. He saw me communicating with my parents with some signs. He made a private decision to teach me Hebrew, which is easier than English. There aren't so many rules and exceptions." By five, Weiner says, he was using ASL, English, Yiddish, some German, and some Hebrew. He and his grandfather spent Wednesday and Thursday evenings going through the Talmud and "talking about philosophy, physics, math, everything." Weiner wasn't challenged in most of the deaf schools he attended and ended up at a public high school that had a resource classroom for deaf students. "During class time, I stuck comic books inside my textbooks. I got nothing from lectures; I succeeded by reading." Until she saw him doing his calculus homework, his grandmother assumed he would learn a trade like most deaf men, says Weiner. "At her funeral, the rabbi said the proudest moment of her life was seeing me go to college."

Originally an engineering student at Hofstra University, Weiner gave that up and moved to California, where what is now the California State University, Northridge, deaf education program was getting off the ground. There, he discovered the perils of poor interpreting. "One time, there was a word in a biology class: 'phosphorylation.' I remember because the interpreter didn't spell the entire word, just P-P-T," he says. "I stopped and said, 'Wait, what's the word?' The professor was angry at me. Half of the next exam was on that word." He routinely fell asleep in a class on Shakespeare until the usual interpreter fell sick and a substitute arrived who used to teach at Gallaudet. "I was, like, 'Holy Day!'" he says. "I became a very important part of the class that week. I was able to express myself fully. I learned that the other students weren't that hot. The teacher said, 'Steve, you have comments?' And I said, 'Yes, I do.'" The experience turned his sense of the world on its head. "My mother and father were taught that hearing people knew more, and they taught me that, too. It wasn't until I could see for myself that I knew that's not always true." When his regular interpreter returned to the English class and things went right back to the way they had been, Weiner threw up his hands. "I said, 'That's it, I'm going to Gallaudet.'"

He came in 1974 as a sophomore, and the higher level of communication was eye-opening. "I was able to meet classmates who understood what I was thinking," he says emphatically. But the professors were a different story. Predominantly hearing or late-deafened, many professors had low opinions of the students, he says. "They were very paternalistic." A prejudice in favor of those with strong verbal skills coursed through the university, a situation Weiner likens to the light skin/dark skin divide in the black community in earlier times. Howard University, a historically black school, is only a few miles up the road from Gallaudet. "The lighter your skin at Howard, the better off you were," says Weiner, noting that when he saw a photograph of Mordecai Johnson, who served as Howard's first black president from 1926 through 1960, he thought he was white. In the deaf community, he maintains, "we still have that problem."

Weiner found mentors in the handful of deaf professors who really inspired him, like a World War II refugee who earned a master's in history at Georgetown University after arriving in the States. In her history class, he says, "she wasn't interested in having us learn dates. We talked about cause and effect, about the abstract not the tangible. She can't talk worth a damn, but she knew inside each one of us we could succeed if expectations were set high enough."

Weiner has high expectations for Gallaudet. He locates the roots of the problems that led to the university's probation in the conflicting messages students like him received there. It was at once a place of inspiration and community but also a somewhat limiting box where the bar of achievement was low. "The 1988 protest was a watershed event," he says. It was also "the moment when we could have changed things." By the early 2000s, however, it was clear that not much had changed beyond the fact that the school had a deaf president. "There were students who had no business being here—they couldn't graduate—and there were professors with low standards," says Weiner. "By the time the 2006 protests began, our dirty laundry was out for everyone to see."

His first day as provost in July 2007 was the day that the Middle States Commission on Higher Education put the university on probation. Choosing to see it as an opportunity to do things differently, Weiner dedicated himself to changing the ethos. The current president of Gallaudet, T. Alan Hurwitz, came in 2010 from Rochester, where he was president of the National Technical Institute for the Deaf. He is very highly regarded but was recovering from surgery when I was on campus, so we didn't meet. Together, Weiner and Hurwitz and their colleagues have tightened standards by requiring higher scores on the ACT standardized admissions test; brought in new staff, including what Weiner calls "top-of-the-line" professors from places like MIT, Harvard, and Columbia; and instituted evidence-based metrics for measuring success. Much effort has been put into diversity, which at Gallaudet includes hearing status. "About fifteen years ago, during a panel discussion on cochlear implants, I raised this idea that in ten to fifteen years, Gallaudet is going to look different," says Weiner. "There was a lot of resistance. Now, especially the new generation, they don't care anymore." Gallaudet does look different. In addition to more cochlear implants, there are more hearing students on campus, mostly enrolled in graduate programs for interpreting and audiology. Both of these new groups do best if they have or quickly acquire proficiency in ASL.

The hard decisions Weiner has had to make as provost have dented his popularity on campus, but he says, "At least I'm honest about the state we're in." On some fronts, that state is much improved. Before 2007, barely half the freshman class made it to sophomore year. In the past four years, the freshman retention rate has ranged from 70 to 75 percent. The latest strategic plan sets a goal of a 50-percent graduation rate. Enrollment is still lower than it was, but Weiner insists the quality of the student body is higher. One sign of Gallaudet's improving status is that today Weiner sits on accreditation committees to evaluate other universities. "It shows they have trust in us," he says.

"I want deaf students here to see everyone as their peers, whether they have a cochlear implant or are hard of hearing, can talk or can't talk. I have friends who are oral. I have one rule: We're not going to try to convert one another. We're going to work together to improve the life of our people. The word 'our' is important. That's what this place will be and must be. Otherwise, why bother?"

Not everyone on campus agrees with him. "We still have what you might call the Republicans, the Democrats, and the Libertarians," he says. "Some faculty say we should all be deaf, others say we need a mix. Sometimes I agree, sometimes I disagree, but I certainly enjoy the diversity of opinions. The day any university becomes groupthink is the day America ceases to function as a real community."

At the end of our time together on my first visit, Weiner hopped up to shake my hand.

"I really want to thank you again for taking time to meet with me and making me feel so welcome," I say.

"There are people here who were nervous about me talking to you," he admits. "I think it's important to talk."

So I make a confession of my own. "I was nervous about coming to Gallaudet as the parent of a child with a cochlear implant," I say. "I didn't know how I'd be treated."

He smiles, reaches up above his right ear, and flips the coil of a cochlear implant off his head. I hadn't realized it was there, hidden in his brown hair. Our entire conversation had been through the interpreter. He seems pleased that he has managed to surprise me.

"I was one of the first culturally Deaf people to get one."

Perhaps it's not surprising that most of the people who talk to me at Gallaudet turn out to have a relatively favorable view of cochlear implants. Irene Leigh doesn't have one, but she is among the Gallaudet professors who have devoted the most time to thinking about them. When we met, she was about to retire as chair of the psychology department after more than twenty years there. A successful product of oral deaf and then mainstream education, Leigh's ability to cope in the hearing world was evident at an early age. Her parents were German refugees who immigrated to the United States a few years after the end of World War II, and her mother made the mistake of mentioning to the immigration inspector on the ship that her daughter was deaf. After an anxious night in detention at Ellis Island, Leigh's parents asked her to draw a tree and other childish subjects for the officials when their case was presented the next day. She followed her parents' spoken instructions beautifully, even though her father's German accent was so thick "you could cut it with a knife." The immigration officer allegedly exclaimed: "My four-year-old grandson can't do that!" And she was in. "I tell people my deafness got me into Ellis Island and it got me out."

For years, Leigh worked at Lexington School for the Deaf as a teacher, then as a counselor and therapist, earning a master's in counseling and a doctorate in clinical psychology along the way. She began signing in her twenties (and got her first interpreter in 1980, when she started her PhD program) and arrived at Gallaudet's psychology department in 1991. Her interest in cochlear implants came early.

"I tend to be a very modern, flexible person," she tells me. "In New York after I got my doctorate, I became a consultant for a cochlear implant program in Manhattan. I got very interested in why people were so opposed." At Gallaudet, she and sociology professor John Christiansen teamed up in the late 1990s to (gingerly) write a book about parent perspectives on cochlear implants for children; it was published in 2002. Immersed as they both were in Gallaudet's culture, they were sensitive to all the issues. At that time, she says, "A good number of the parents labeled the Deaf community as being misinformed about the merits of cochlear implants and not understanding or respecting the parents' perspective." For their part, the Deaf community at Gallaudet was beginning to get used to the idea, but true supporters were few and far between. In 2011, Leigh served as an editor with Raylene Paludneviciene of a follow-up book examining how perspectives had evolved. By then, culturally Deaf adults who had received implants were no longer viewed as automatic traitors, they wrote. Opposition to pediatric implants was "gradually giving way to a more nuanced view." The new emphasis on bilingualism and biculturalism, says Leigh, is not so much a change as a continuing fight for validation. The goal of most in the community is to establish a path that allows implant users to still enjoy a Deaf identity. Leigh echoes the inclusive view of Steve Weiner when she says, "There are many ways of being deaf."

When I knocked on the open door of Sam Swiller's office, he saw me rather than heard me. Engrossed in his computer and a pile of architectural plans, he looked up, said hello, and reached up to turn on his cochlear implant. Like his brother Josh, Swiller is not deaf in the expected Gallaudet way. He grew up speaking and listening with hearing aids and didn't get a cochlear implant until he was twenty-nine. "The absolute clarity of a windshield wiper blew my mind for days," he says with a laugh. Born in 1975, he was already nearly a teenager when the clinical trials for cochlear implants began. His parents thought the technology crude and were suspicious because the strongest proponents seemed to be surgeons, who had something to gain from promoting a surgery.

Growing up with hearing loss made the normal teenage difficulties worse, says Swiller. "You lose your personality a little bit because you're so focused on making sense of what's being said to you. You have to work harder; it's stressful. The excitement of speaking to a cute girl, it gets magnified. Am I pronouncing my 's' right? Am I spitting on her?"

He finally got the implant after his residual hearing deteriorated dramatically. "With hearing aids, I was able to hear maybe a third of what was spoken, then fill in the blanks," he says. "The implant boosted that to about 60 percent. Now I'm aware more than ever of the 60 percent I missed. It's given me a huge boost in confidence and a little bit more of a boost in hearing. The CI is what I'm doing for myself to help myself. Now I have it, I think it's amazing. Life is difficult and you need every type of weapon in your quiver, every resource possible. It's not a solution for everyone. I'm not trying to put my ideas on anyone else."

Even with his cochlear implant, he has found a professional home at Gallaudet, and much-needed comfort and ease. Before he got there,

Swiller went through a difficult period in both his professional and personal life. "I was questioning my strategy of trying to find work in a very competitive financial field where, whether or not they are real, I'm perceived to have strikes against me," he says of his hearing loss. Following his older brother's example, he came to Gallaudet, initially as a visiting professor in the business department. It was challenging on many levels, and the "hardest part by far was becoming proficient in ASL: I had to have an interpreter." Still, he found himself enjoying this new environment. "I felt the students were giving me a lot and I was gaining a new perspective on what deafness was. It was a celebration and it was beautiful. In the past, I felt proud of what I'd overcome. But coming here, I was really celebrating it and growing in that regard." He even found that little quirks he'd thought grew out of his own anxiety—like looking over his shoulder all the time—were common in everyone around him.

Now working as a vice president managing real estate for the university, Swiller is helping to open up the school to the outside world. For decades, the surrounding neighborhood in northeast Washington was dangerous enough to keep students mostly within the gates. But it's an area in the midst of a rebirth, especially along what's known as the H Street corridor, and Gallaudet happens to own real estate there. "The wall between the hearing world and the deaf world is getting shorter. It's a more porous border," says Swiller. He was speaking metaphorically, but it could have been literal, too. "The students of your son's generation are going to be able to [cross] that border."

When I left Swiller, I went to see Matthew Bakke, who heads Gallaudet's Department of Hearing, Speech, and Language Sciences. Unlike everyone else I met at the university, Bakke is hearing, but he has been involved with deaf people his whole life. Growing up in the 1950s, he had a younger brother who was deaf. "He ended up essentially being a marginalized person because of his education and experiences," says Bakke. "He went away to residential school at the age of three. Monday morning to Friday evening, he was gone. His education was oral even though he is profoundly deaf. A hearing aid did him almost no good. That was what was available." After stints in the seminary and the army, Bakke gravitated, like a lot of siblings of deaf people, to being a teacher of the deaf and then an audiologist. His teaching experiences taught him the importance of beginning very early with children. When the children didn't get language early, he says, "the education was essentially futile. By the time I got them, they had missed so much."

Initially, he was deeply skeptical that cochlear implants could possibly work. When he saw the video of Bill House and Jack Urban activating an early implant for Karen, the deaf young woman who listened to Beethoven, Bakke's reaction was not wonder but disgust. "I thought, 'This is just not reality.' I felt they were exploiting her and it made me sick." Then he went to a conference where he met a group of children with cochlear implants. "I saw what they were doing," he says. "It was a Road to Damascus conversion. I had worked on oral English speech development. I knew how difficult it was. These kids weren't doing that great compared to what we see now, but it was orders of magnitude better than at the time. I said, 'These work.'"

That doesn't mean he thinks deaf kids don't need ASL. "The message [at Gallaudet] is: Don't deprive deaf children of sign language, and I don't disagree," he says. "I support children being given everything possible. It's destructive to say that ASL is going to interfere. That's not based on science, that's based on bias. It's the same as with the deaf saying don't [get an implant]—that's a cultural bias. But I have a feeling that to survive or prosper really in our world and our culture, English is required."

After decades working in the field, Bakke now has a personal relationship with cochlear implants. His granddaughter, who is still a toddler, has two. Her parents have signed with her since they knew she was deaf. Today, the little girl signs and speaks, though her spoken language is still somewhat delayed. "I do understand why people would feel hurt about a societal perspective that there's something wrong with you," says Bakke. "Is 'fixing' the right word? My grandchild did not need to be fixed. She's a wonderful human being and a gift to the world. She can't hear. I don't consider it fixing. I consider it giving her access. When you give a child access, you give a child the ability to learn the way other people are able to learn."

Views like Bakke's are still not popular at Gallaudet, and he walks a delicate line there. "Because of my brother, I feel very at home here," he says. "I had a foundation in ASL. I felt more comfortable coming here than a lot of people would." Nonetheless, he sees Gallaudet's future as "uncertain." His own view of the new realities of deafness is clear-eyed. "The cochlear implant is the best thing that ever happened to deaf kids," he says. "But it isn't a miracle or a cure. It isn't even very good. It has real limitations. It's a wonderful device that opens the world for kids, but it's not hearing. They're not hearing kids. They're deaf kids." Even so, the change implants have brought is extreme. Says Bakke, "My brother's life was A and my grandchild's life is Z."

Janis Cole, my ASL teacher, does not have a cochlear implant, nor does she want one. I guessed from an early class discussion on Deaf culture and appropriate labels that she was suffused with Deaf pride and contentment. So I leaked out the details of Alex's life over the two weeks we were together.

"Thank you for sharing about your son," she wrote after I turned in a homework assignment in which I asked what the difference between deaf and hard of hearing meant to her (without letting on about the implant). She suggested we get together with an interpreter.

The schedule didn't work in our favor, and when there was no interpreter to be had, she said we should meet after class alone anyway. "We can do it!" she said.

We sat in the classroom one afternoon after everyone was gone, communicating by signing and writing on a piece of paper. I still have the paper. It's a collage of emotion. At the center is a list of resources to help me access the deaf community in New York, including a good friend's e-mail address. But all around the margins of that sheet of lined notebook paper, there are signs of an intense conversation.

I WAS SCARED TO COME HERE, I signed at the beginning. MY SON HAS A COCHLEAR IMPLANT.

I GUESSED, she responded.

On the paper, she drew an arrow and wrote the words "extreme" at one end and "liberal" at the other. "Like in any community," she added. I circled the word "extreme" and wrote below it, "so loud."

"They mean well," she wrote on the opposite margin. "OPPRES-SION."

Smack in the middle, she wrote "Bilingual" and circled it heavily. ENGLISH, she signed. Then she fingerspelled A-S-L.

"Literacy," I wrote. And "worried."

We debated that a little, with me signing BOOK and READ and WRITE. In another spot, she wrote "Both," meaning, Why not have Alex do both English and ASL?

The name AG Bell is scribbled across the left-hand corner of the page. Janis made it clear how much she disliked the listening and spoken language advocacy group. I wasn't brave enough to tell her the reason I had to leave class a day early was to collect Alex from my mother and fly to Phoenix for an AG Bell convention, but I did say that I thought they had helped a lot of people.

Then she wrote a longer passage: "Many culturally deaf people can hear and speak yet chose Deaf world. Not necessarily mean they not integrate in the hearing . . ."

She considered for a moment. HEARING, she signed, rolling her forefinger away from her mouth. Then, a finger to her ear and her mouth in succession: DEAF.

"Walk in front of me," she wrote. Then she put an X through "front." And wrote underneath it: "Walk beside me."

It was a quote usually attributed to Albert Camus: "Don't walk behind me, I may not lead. Don't walk in front of me, I may not follow. Just walk beside me and be my friend." It seemed a fair enough piece of advice to guide the interaction between deaf and hearing.

There was a pause and then, to my great surprise, Janis spoke.

"I'm going to talk," she said very clearly. "Because it's important." Then she paused. "I never use my voice here."

I could see that it cost her something to do so, yet the desire to connect had moved us both. I started to talk and she shook her head.

"You still have to sign," she reminded me.

OF COURSE, I signed, surprised and embarrassed at how unthinkingly I had reverted to my way of communicating.

She looked at me intently and said with feeling: "I think it's wonderful you're here."

THANK-YOU, I signed, touching my chin with four fingers and then moving my hand toward her. I'M GLAD TO MEET YOU.

Then I wrote it down for emphasis. Under "meet" I added "know": I am glad to know you.

We stood up. We saw the world a little differently, but that was okay.

27

FROG IN HONG KONG

Mark had something on his mind. He had gotten home late from work. The boys were in bed and we were catching up over a glass of wine. He had a tentative look on his face as if he'd hit pause somewhere between wishful and worried. Then he blurted it out.

"Would you consider moving to Hong Kong?"

I stared at him and tried not to overreact.

"What?!"

"They've asked me to move to Hong Kong."

For a long minute, I didn't say anything.

"You're not saying no," he pointed out, surprised.

I was thinking. I really didn't want to move to Hong Kong. I had dug my roots deep into the tree-lined streets of Brooklyn. But I knew Mark wanted to go. He and I had lived in Europe for six years and had loved the experience. He had always wanted to live in Asia, too. Plus he wanted—needed—a change at the bank where he worked. Mostly, though, I wasn't thinking about Mark or about me.

"What about Alex?" I finally said.

Alex was halfway through first grade, and I had finally taken a deep breath and dared to think we were over the highest hurdles. He was reading—phonics-heavy early readers featuring sequences like "the hen is mad, the hen is in the mud, the mad hen is wet" and early Dr. Seuss, like *Hop on Pop*. He had friends—lovely friends who stopped to place his processor back on his head if the magnet fell off in the playground, which it frequently did. It didn't seem fair to take him somewhere where it might be considerably harder for him to thrive. We had a carefully cultivated support network: a school we loved, a top-notch doctor, a teacher of the deaf, a speech therapist, and three— yes, three—audiologists (one for the implant, one for the hearing aid, one from the Board of Education to work with his school). Did we need all that? Could we take him halfway around the world and start all over again? Did his success depend on his particular situation or did it mean he was ready to move beyond it?

What weight did his hearing carry relative to all the other consid-erations that determine life decisions: Mark's career versus mine, the cost of this versus the savings from that, the happiness of this kid versus the education of that kid, change versus status quo, thirst for adventure versus desire for rootedness? Alex's needs counted for a lot. But he really was doing awfully well. I opened the door to a small thought: It might not be impossible. The idea was simultaneously frightening and freeing.

A few nights later, we sat the boys down and told them what we were considering. Jake, who was eleven, was teary and shocked. He acknowledged that there might be cool aspects to such an adven-ture, but he had specific questions like: "Will the pictures be in the same spots on the walls when we come back?"

Matthew, at eight, was dead set against it: "No! No! No! You can't make me go."

Alex, who was six, could barely contain himself. Gleefully, he cried: "We're moving to China!"

Then he flung himself onto Mark's lap, wrapped his arms around his neck, and declared: "You're the best parents ever!"

Speechless, Mark and I just stared at each other.

"Okay, I wasn't expecting that," I said finally.

"Wow," was all Mark could manage.

Jake and Matthew started to protest. "He's only six. He doesn't understand what it means to leave his friends."

"I'll Skype," said Alex confidently, which made us all laugh, since not one of us had ever used Skype. I guessed he had talked to our German babysitter's friend back home. Alex turned back to Mark: "I want to see the Great Wall. I want to learn Mandarin, more than just *Ni hao*. I'll even try some Chinese food!"

"What does he know?!" cried Jake and Matthew in unison.

They had a point. Alex couldn't possibly grasp what it truly meant to uproot our family and move around the world to a very foreign culture. Neither could he think through how much more difficult it might be for him in particular to find a school that worked, therapists to help him continue his progress, and a community that was as warm and nurturing as the one we had in Brooklyn. There would be even fewer people with hearing aids and cochlear implants in his life.

But Alex inspired me. "Let me go!" he was saying. He was brave and adventurous; I should be, too.

I got on a plane to investigate what life in Hong Kong would look like. It wasn't all encouraging. There was no such thing as a teacher of the deaf. One speech therapist I interviewed had such a heavy Chinese accent it was hard even for me to understand her. We would have to buy some expensive new equipment, since our American FM system— the microphone the teacher used in the classroom—wouldn't work in Hong Kong. And who would fix a broken hearing aid or processor? That last question was no small matter. In the space of one frustrating two-year period, Alex's cochlear implant processor was repaired fifteen times. One day, his hearing aid had broken, too, and I had had to leave him standing in the classroom, still and staring, with the same expression he had worn years earlier at that first evaluation. His success was undeniably dependent on technology.

Not long after my first trip to Hong Kong, on the weekend of Alex's seventh birthday, I took him and his brothers to Boulder, Colorado, to meet neuroscientist Anu Sharma. She was going to include Alex in her database of deaf and hard-of-hearing children. Jake and Matthew were there to serve as typically hearing children for the study and as moral support for Alex. The EEG measurements Sharma would take of Alex would allow us to compare the speed of his responses to acoustic signals on both sides of his brain, through the hearing aid and through the cochlear implant. And they would help her in her quest to understand if there were clues in the brain to tell us which kids were going to do better with cochlear implants than others. "Say one hundred children get implanted at one year," says Sharma. "They're not all like Alex. Some won't speak that well or hear that well. What about the brain causes that?"

It was a beautiful April day when we drove to the laboratory where Julia the lab manager met us and showed us the fancy cap of electrodes they had recently begun using. Alex started out brave.

"I want to go first because I'm the one with this," he said, pointing to his processor.

Once in the soundproof booth, though, sitting alone in the big black vinyl chair as lab staff wielding wires surrounded him, he began to look nervous and called for me to come in and sit with him. Jake and Matthew would use the new "net" of electrodes, a cap that was easy to put on and take off. For Alex, Julia used a conductive gel and taped nine electrodes to various spots on his head: one at the top, two behind and below each ear, and the rest around his forehead and on the top of his nose.

On a television in front of him, *The Chronicles of Narnia: The Lion, the Witch, and the Wardrobe* played silently. Suddenly, "ba, ba, ba" blared from the speakers to Alex's left and right. "It sounds like a frog," said Matthew later, when it was his turn. Alex squeezed my hand and together we listened to the full-throated croaking—ba, ba, ba—and watched Lucy meet Mr. Tumnus by the lamppost in the snowy woods of Narnia.

Just like they were supposed to, the repeated speech sounds evoked a response in Alex's brain. The results—line after line of waveforms—brought good news. There was a range for normal and Alex was in it. Through both ears, with the hearing aid and the cochlear implant, his auditory cortex had been consistently receiving enough sound that it had developed an N1 and P1 that could have been that of a hearing child.

As if to prove it, that same night, at a noisy Mexican restaurant with friends who lived in Boulder, Alex insisted we play Whisper down the Line, his new favorite game. He was terrible at it, mangling the words that I passed on to him after Matthew passed them on to me. But still . . . who would have ever thought? He could hear me whisper.

Nearly a year after Mark suggested we move, Alex and I sat at the table in our new apartment with his second-grade homework from the Hong Kong International School.

I had double-checked the batteries in Alex's cochlear implant and hearing aid, so I knew that when he first ignored my request to sit down and focus, it was probably not because he didn't hear me but because he wanted to finish telling Matthew about the funny thing that happened on the school bus. After one more trip to the kitchen for another yogurt, though, he had settled down to show me what he'd learned about how to complete the double- and triple-digit arithmetic problems he'd been assigned.

When we moved on to reading, which he had to do for twenty minutes, he pulled out his beloved copy of *Frog and Toad All Year* and picked a story to read aloud. Since the holidays were coming, he chose "Christmas Eve," in which Frog is late for Christmas Eve supper and Toad imagines all the horrible things that might have befallen his friend.

"What if Frog has fallen into a deep hole and cannot get out?" read Alex.

"'What if Frog is lost in the woods?'" He added some drama to his voice. "'What if he is cold and wet and hungry? What if he is being chased by a big animal with many sharp teeth? What if he is being eaten up?'" cried Alex.

Fortunately, Frog had simply gotten delayed wrapping Toad's present. Alex turned reflective as he read the last lines about the two friends enjoying a quiet evening by the fire. He closed the book.

"I can't wait for Christmas, Mommy," he said. "When can we get our tree?"

We talked about Christmas for a few minutes—it would be our first away from family and home.

Then I reminded him to enter the book's title into his reading log. There was a spot for assessing the book's difficulty. A month before, he had called it "Just Right."

"I think it's 'Too Easy' now," he admitted.

I gave him a hug, a kiss, and a "good job, sweetheart," and he went off to play with his new friend Christopher who lived upstairs.

Except for the bit about the batteries, it could have been a scene at anyone's dining room table.

NOTES

Comments that came from my interviews are indicated in the present tense ("says" rather than "said"). Those from other sources are cited below.

CHAPTER 1: THE COW AND THE RED BALLOON

For personal chapters, I relied on my journals, recollections of events, and files of reports about Alex from audiologists, doctors, speech therapists, and teachers. I also interviewed some of the professionals who have worked with Alex over the years.

5 **David Kemp discovered:** Information on otoacoustic emissions (OAEs) is available at http://www.asha.org/public/hearing/Otoacoustic-Emissions/. OAEs and other diagnostic hearing tests are also described in Debby Waldman with Jackson Roush, *Your Child's Hearing Loss: What Parents Need to Know* (New York: Perigee, 2005). An American Academy of Audiology interview with David Kemp is at http://www.audiology.org/news/Pages/20090106b.aspx.
Arlene Eisenberg, Heidi Murkoff, and Sandee Hathaway, *What to Expect the First Year* (New York: Workman, 1996). [Latest edition published 2010.]

CHAPTER 2: A NEW WORLD

14 **Two or three in a thousand:** For statistics on hearing loss, I referred to the National Institute on Deafness and Other Communication Disorders (NIDCD), at http://www.nidcd.nih.gov/health/statistics/Pages/quick.aspx; the Centers for Disease Control (CDC), at http://www.cdc.gov/ncbddd/hearingloss/data.html; and the American Speech-Language-Hearing Association, at http://www.asha .org/aud/Facts-about-Pediatric-Hearing-Loss/.

15 **Sam Supalla:** In Carol Padden and Tom Humphries, *Deaf in America: Voices from a Culture* (Cambridge, MA: Harvard University Press, 1988), 15–16. This story was also recounted to me by Ted Supalla. Also in Padden and Humphries, *Deaf in America*: **"For hearing people"**, 92; **a different center**, 41; **"A Deaf couple,"** 103; THINK-HEARING, 53.

16 **deaf community's pride:** To read about identity and illness, see chap. 1, "Son," in Andrew Solomon, *Far from the Tree: Parents, Children, and the Search for Identity* (New York: Simon & Schuster, 2012).

18 **You could be deaf or Deaf:** There is a full discussion of terms used to describe hearing loss, varieties of hearing loss, and cultural terms in the first two chapters of Marc Marschark, *Raising and Educating a Deaf Child* (New York: Oxford University Press, 2007), and a thoughtful discussion on this subject in "Identity and the Power of Labels," the first chapter of Irene W. Leigh, *A Lens on Deaf Identities* (New York: Oxford University Press, 2009). The National Association of the Deaf gives a full explanation of its views on labels at http://www.nad.org/issues/american-sign-language/community-and-culture-faq. See also AG Bell's listening andspokenlanguage.org.

19 **a tribe in Namibia:** See "Do You See What I See?" BBC Two *Horizon*, at http://www.bbc.co.uk/news/science-environment-14421303.

20 **"Deafness as such":** From Oliver Sacks, *Seeing Voices* (New York: Vintage, 1989), 94.

20 **"Language really does":** Paula Tallal's interview on Phonological Processing, with David Boulton, Children of the Code, at http://www.childrenofthecode .org/interviews/tallal.htm.

21 **"equally suitable for making love or speeches":** Sacks, *Seeing Voices*, 101. For discussion of deaf educational underachievement, see Marschark, *Raising and Educating a Deaf Child*, 165; John B. Christiansen, *Reflections: My Life in the Deaf and Hearing Worlds* (Washington, DC: Gallaudet University Press, 2010) 214; Commission on Education of the Deaf, *Toward Equality: Education of the Deaf*, Report to the President and the Congress of the United States, 1988; Sue Archbold and Gerard M. O'Donoghue, "Education and Childhood Deafness: Changing Choices and New Challenges," in John K. Niparko, ed., *Cochlear Implants: Principles & Practices*, 2nd ed. (New York: Lippincott Williams & Wilkins, 2009). For deaf reading levels, see Carol Bloomquist Traxler, "The Stanford Achievement Test, 9th Edition: National Norming and Performance Standards for Deaf and Hard-of-Hearing Students," *Journal of Deaf Studies and Deaf Education* 5 no. 4, (2000): 337–348; Marc Marschark and Patricia Elizabeth Spencer, "Epilogue: What We Know, What We Don't Know, and What We Should Know," in *The Oxford Handbook of Deaf Studies,*

Language and Education, vol. 1, 2nd ed. (New York: Oxford University Press, 2011).

Deaf graduation and employment rates: Blanchfield BB, Feldman JJ, Dunbar JL, Gardner EN. The severely to profoundly hearing-impaired population in the United States: "Prevalence Estimates and Demographics," *Journal of the American Academy of Audiology* (2001): 12:183–189. National Technical Institute for the Deaf Collaboratory on Economic, Demographic, and Policy Studies: http://www.ntid.rit.edu/research/collaboratory.

22 **"a miracle of biblical proportions":** Steve Parton, quoted in Edward Dolnick, "Deafness as Culture," *Atlantic Monthly*, September 1993.

22 **National Association of the Deaf . . . position on cochlear implants:** NAD position papers, 1990, 2000. See also: John B. Christiansen and Irene W. Leigh, *Cochlear Implants in Children: Ethics and Choices* (Washington, DC: Gallaudet University Press, 2002), 257.

23 **"genocide" and "child abuse":** See Dolnick, *Atlantic Monthly*; Andrew Solomon, "Defiantly Deaf," *New York Times Magazine*, August 28, 1994.

CHAPTER 3: HOW LOUD IS A WHISPER?

25 **"maniacal miniature golf course":** Diane Ackerman, *A Natural History of the Senses* (New York: Vintage, 1991), 177.

25 **our keenest hearing:** E. Bruce Goldstein, *Sensation and Perception*, 8th ed. (Belmont, CA: Wadsworth Cengage Learning, 2010), 265; Thomas D. Rossing, F. Richard Moore, and Paul A. Wheeler, *The Science of Sound*, 3rd ed. (Reading, MA: Addison-Wesley, 2002), 335.

25 **Prehistoric ears:** Discussed in section on ear damage in Rossing et al., *Science of Sound*, 722.

For sound waves, the mechanics of hearing, and decibel levels, see Rossing et al., *Science of Sound*, chaps. 1 and 5; Goldstein, *Sensation and Perception*, 261–272; House Clinic, *The Complete Idiot's Guide to Hearing Loss* (New York: Alpha Books, 2010), 10–13.

For sound versus vision, see Rossing et al., *Science of Sound*, 79–80.

29 **more going on than that:** Daniel J. Levitin, *This Is Your Brain on Music: The Science of a Human Obsession* (New York: Dutton, 2006), 41–46.

For frequency of speech sounds/speech banana, see American Academy of Audiology table, "Audiogram of Familiar Sounds."

For bone conduction, see Rossing et al., *Science of Sound*, 85. ABR is described in Waldman and Roush, *Your Child's Hearing Loss*, 34.

32 **In one state's survey:** Marschark, *Raising and Educating a Deaf Child*, 14–15, citing research by Christine Yoshinaga-Itano.

CHAPTER 4: A STREAM OF SOUND

34 **nearly seven thousand languages:** http://www.ethnologue.com/world.

34 **Most begin to talk:** See Charles Yang, *The Infinite Gift: How Children Learn and Unlearn the Languages of the World* (New York: Scribner, 2006), 2–3; and Roberta Michnick Golinkoff and Kathy Hirsh-Pasek, *How Babies Talk: The Magic and Mystery of Language in the First Three Years of Life* (New York: Dutton, 1999).

34 **"doubtless the greatest intellectual feat":** Leonard Bloomfield, *Language* (New York: Henry Holt, 1933).

34 **From Saint Augustine to Charles Darwin:** Alison Gopnik, Andrew N. Meltzoff, and Patricia K. Kuhl, *The Scientist in the Crib: Minds, Brains, and How Children Learn* (New York: Morrow, 1999), 98.

35 **Chomsky disagreed:** For Skinner versus Chomsky, see Yang, *Infinite Gift*, 16–18.

35 **Universal grammar explained:** Yang, *Infinite Gift*, 8, 16–31; Gopnik et al., *Scientist in the Crib*, 99–102.

36 **Victor of Aveyron:** Harlan Lane, *When The Mind Hears: A History of the Deaf* (New York: Vintage, 1989), 122–132.

36 **Genie:** Susan Donaldson James, "Wild Child 'Genie': A Tortured Life," Associated Press, May 8, 2008, at http://abclocal.go.com/wpvi/story?section=news/national_world&id=6130233; Gopnik et al., *Scientist in the Crib*, 192.

37 **"miniature languages":** See Maryia Fedzechkina, T. Florian Jaeger, and Elissa L. Newport, "Language Learners Restructure Their Input to Facilitate Efficient Communication," *PNAS* 109 no. 44 (2012): 17897–17902; and Carla L. Hudson Kam and Elissa L. Newport, "Getting it right by getting it wrong: When learners change languages," *Cognitive Psychology* 59 no. 1 (2009): 30–66. For discussion of the "less is more" theory, see also Elissa L. Newport, "Maturational Constraints on Language Learning," *Cognitive Science* 14 no. 1 (1990): 11–28.

39 **Hart and Risley:** Betty Hart and Todd R. Risley, *Meaningful Differences in the Everyday Experience of Young American Children* (Baltimore: Paul H. Brookes Publishing, 1995). Quotes from Risley come from an interview by David Boulton for Children of the Code, at http://www.childrenofthecode.org/interviews/risley.htm.

41 **a lively, ongoing debate:** Interview with Elissa Newport. See also Elissa L. Newport, "Plus or Minus 30 Years in the Language Sciences," *Topics in Cognitive Science* 2 no. 3 (2010): 367–373; and Elissa L. Newport, "The Modularity Issue in Language Acquisition: A Rapprochement? Comments on Gallistel and Chomsky," *Language Learning and Development* 7 no. 4 (2011), 279–286.

41 **"nurture *is* our nature":** Gopnik et al., *Scientist in the Crib*, 8.

42 **"Children . . . are innovators" . . . Songbirds interest linguists:** Yang, *Infinite Gift*, 4–5.

45 **the somewhat methodical way babies focus:** Interviews with Athena Vouloumanos and with Gina Lebedeva, director of Translation, Outreach, and Educa-

tion, Institute for Learning and Brain Sciences, University of Washington (lab of Patricia Kuhl). See also: Judit Gervain, Iris Berent, and Janet F. Werker, "Binding at Birth: The Newborn Brain Detects Identity Relations and Sequential Position in Speech," *Journal of Cognitive Neuroscience* 24 no. 3 (2012): 564–574; Athena Vouloumanos and Janet F. Werker, "Listening to Language at Birth: Evidence for a Bias for Speech in Neonates," *Developmental Science* 10 no. 2 (2007): 159–171; Athena Vouloumanos and Janet F. Werker, "Tuned to the Signal: The Privileged Status of Speech for Young Infants," *Developmental Science* 7 no. 3 (2004): 270–276; Athena Vouloumanos et al., "Five-month-old infants' identification of the sources of vocalizations," *PNAS* 106 no. 44 (November 3, 2009): 18867–18872; Gopnik et al., *Scientist in the Crib*, 106–110.

CHAPTER 5: "SOME MEANS OF INSTRUCTING"

My main sources on the early history of deaf education are Harlan Lane, *When the Mind Hears: A History of the Deaf*; David Wright, *Deafness: An Autobiography* (New York: HarperPerennial, 1993); and Carol Padden and Tom Humphries, *Deaf in America*, and their second book, *Inside Deaf Culture* (Cambridge, MA: Harvard University Press, 2005). Page references below as well as additional sources.

50 **Abbé Charles-Michel de l'Epée:** Quotes are from Lane, *When the Mind Hears*, 57–58. See also Padden and Humphries, *Deaf in America*, 27—29, and Wright, *Deafness*, 183.

52 **"It has come to symbolize":** Padden and Humphries, *Deaf in America*, 29.

52 **Samuel Johnson:** From *A Journey to the Western Islands of Scotland*, quoted in Wright, *Deafness*, 179.

53 **indistinguishable from that of the mentally ill:** Wright, *Deafness*, ix.

53 **Ponce de Léon:** Ibid., 164–168, and Lane, *When the Mind Hears*, 90–94. Pedro de Velasco is quoted in both books. I have used the translation from Wright.

54 **"a breakthrough that shattered":** Wright, *Deafness*, 171.

54 **Juan Pablo Bonet:** Ibid., 167–171, and Lane, *When the Mind Hears*, 86–94.

54 **Epée's system:** Wright, *Deafness*, 183–186; Lane, *When the Mind Hears*, 36, 58–63; Marschark, *Raising and Educating a Deaf Child*, 71.

54 **Epée prevailed upon the French state:** Wright, *Deafness*, 187–188; Lane, *When the Mind Hears*, 33.

55 **Abbé Roch-Ambroise Cucurron Sicard:** Lane, *When the Mind Hears*, 32–41; Wright, *Deafness*, 188.

55 **twelve more similar schools . . . rose to sixty:** Lane, *When the Mind Hears*, 64.

55 **Jean Massieu:** Ibid., 17–23.

55 **"What is hope? . . . vigor of the mind":** Ibid., 22–23.

55–56 **Johann Conrad Amman . . . and John Wallis:** Ibid., 100–106.

56 **"The breath of life":** Ibid., p. 100. Wright, *Deafness*, 172–178.

56 **Samuel Heinicke:** Wright, *Deafness*, 186–187; Lane, *When the Mind Hears*, 102–103 (taste technique, 103); Gabriel Grayson, *Talking with Your Hands, Listening with Your Eyes* (Garden City Park, NY: Square One, 2003), 3. Alexander Graham Bell Association website: www.listeningandspokenlanguage.org (the history page is no longer part of the website).

57 **Heinicke sent Epée an extensive argument:** Wright, *Deafness*, 186.
Oral versus manual: Ideas mentioned here are recounted in many places and were discussed in many interviews I did, but the early arguments are summarized in Wright, *Deafness*, xv–xvii, 207, 226–229, and contemporary arguments in Marschark, *Raising and Educating a Deaf Child*, 90–91, on the website of the American School for the Deaf at http://www.asd-1817.org/page.cfm?p=430, and on the AG Bell website at http://listeningandspokenlanguage.org/Document.aspx?id=387.

59 **Gallaudet paid particular attention to . . . Alice Cogswell:** Lane, *When the Mind Hears*, 173–176, and Edward Miner Gallaudet, *Life of Thomas Hopkins Gallaudet, Founder of Deaf-Mute Instruction in America*, originally published by Henry Holt, 1888, reprinted by Forgotten Books, 2012, 46–57.

59 **"immediate and deep . . . instruct her":** Lewis Weld, Alice Cogswell's brother-in-law, quoted in Gallaudet, *Life of Thomas Hopkins Gallaudet*, 47–48.

59 **Gallaudet set off for Britain:** Lane, *When the Mind Hears*, 185–205; Gallaudet, *Life of Thomas Hopkins Gallaudet*, 57–110.

60 **Oh! how this poor heathen people:** Gallaudet, quoted in Lane, *When the Mind Hears*, 196.

60 **The American Asylum, the first school:** See website of American School for the Deaf at http://www.asd-1817.org/page.cfm?p=429; Lane, *When the Mind Hears*, 222 and 238; Wright, *Deafness*, 197.

60 **Edward Miner Gallaudet:** Lane, *When the Mind Hears*, 276–278; Edward Miner Gallaudet, "A History of the Columbia Institution for the Deaf and Dumb," *Records of the Columbia Historical Society*, 1912.

61 **Mabel Hubbard:** Lane, *When the Mind Hears*, 312–315; Charlotte Gray, *Reluctant Genius: Alexander Graham Bell and the Passion for Invention* (New York: Arcade, 2006), chap. 4; website of Clarke School at http://www.clarkeschools.org/about/welcome.

61 **a grant of $50,000:** Lane, *When the Mind Hears*, 320.

61 **Alexander Graham Bell:** In addition to sources named above, my main sources on Bell were Gray, *Reluctant Genius* (e-book, so chapters given rather than page numbers), and Edwin S. Grosvenor and Morgan Wesson, *Alexander Graham Bell: The Life and Times of the Man Who Invented the Telephone* (New York, Abrams, 1997).

62 **Visible Speech:** Gray, *Reluctant Genius*, chaps. 1–3; Wright, *Deafness*, 212–216.

62 **George Sanders:** Bell quoted in Grosvenor and Wesson, *Alexander Graham Bell*, 40.

63 **Helmholtz's 1863 book:** *On the Sensations of Tone as a Physiological Basis for the Study of Music,* 4th ed., trans. A. J. Ellis (Mineola, NY: Dover, 1954).

63 **Bell became convinced:** Grosvenor and Wesson, *Alexander Graham Bell,* 47.

63 **"Mr. Watson, come here":** Ibid., 66; Gray, *Reluctant Genius,* chap. 8.

63 **"He came to his miracle":** Robert Bruce writing in the Foreword to Grosvenor and Wesson, *Alexander Graham Bell.*

63 **at Hartford, Bell learned some sign language:** Gray, *Reluctant Genius,* chap. 3, and Marc Marschark in introduction to "The Question of Sign-Language and the Utility of Signs in the Instruction of the Deaf: Two Papers by Alexander Graham Bell (1898)," *Journal of Deaf Studies and Deaf Education* 10 no. 2 (2005): 111–121.

64 **"Only the intensity":** Gray, *Reluctant Genius,* chap. 3.

64 **His mother initially objected to his marriage:** Ibid., chap. 7.

64 **"When I was young":** Ibid., chap. 8.

65 **his interest in heredity:** Ibid., chap. 12; Lane, *When the Mind Hears,* 353–361; Padden and Humphries, *Inside Deaf Culture,* 174–175.

65 **"He was not as clearly definite":** Marc Marschark in introduction to "The Question of Sign-Language."

66 **conference of deaf educators in Milan:** Lane, *When the Mind Hears,* 387–395; Wright, *Deafness,* 208–209.

66 **deaf students in America being educated in the oral method:** Padden and Humphries, *Inside Deaf Culture,* 48.

66 **"oralist tradition":** Lane, *When the Mind Hears,* 111.

66 **"The weather of the two worlds":** Wright, *Deafness,* 95.

66 **"If knowledge can be compared":** Ibid., 69.

67 **"a commentary":** Ibid., 145.

CHAPTER 6: "MARVELOUS MECHANISM"

68 **Jean Marc Gaspard Itard:** Lane, *When the Mind Hears,* 132–134.

69 **Hermann von Helmholtz:** Rossing et al., *Science of Sound,* 85, 128–129; Harvey Fletcher and H. D. Arnold, *Speech and Hearing* (New York: D. Van Nostrand, 1929), 118–119.

69 **"With one broad sweep":** Fletcher and Arnold, *Speech and Hearing,* xi.

69 **Bell Laboratories:** Jon Gertner, *The Idea Factory: Bell Labs and the Great Age of American Innovation* (New York: Penguin, 2012); S. Millman, ed., *A History of Engineering and Science in the Bell System: Communication Sciences (1925–1980)* (Indianapolis: AT&T Bell Laboratories, 1984) 93–110.

70 **Harvey Fletcher:** Video of interview by Bruce Bogert circa 1963, at http://audi torymodels.org/jba/BOOKS_Historical/FletcherVideo/mpg/fletcher.mpg; Stephen H. Fletcher, *Harvey Fletcher 1884–1981: A Biographical Memoir,* (Washington, DC: National Academy of Sciences, 1992) at http://www.nasonline.org/

publications/biographical-memoirs/memoir-pdfs/fletcher-harvey.pdf (**"all there was to know,"** p. 174); Jont B. Allen, *Articulation and Intelligibility* (San Rafael, CA: Morgan & Claypool, 2005), 24–25; Jont B. Allen, "Harvey Fletcher's Role in the Creation of Communication Acoustics," *Journal of the Acoustical Society of America* 99 no. 4 (1996): 1825–1839.

70 **"accurately describe":** Fletcher and Arnold, *Speech and Hearing*, v.

70 **"The atmosphere of sounds":** Ibid., xi.

71 **"The processes of speaking and hearing":** From Harvey Fletcher, *Speech and Hearing in Communication* (Princeton, NJ: D. Van Nostrand, 1953), 1953 edition of the 1929 book.

71 **"readily interpreted by the eye":** Fletcher, *Speech and Hearing*, 26.

71 **the perfection of this instrument:** Ibid., 26–27.

72 **"farmers"** (vowel sounds, etc.): Ibid., 29–63.

73 **"makes it possible":** Ibid., 49.

73 **After World War II:** *A History of Engineering and Science in the Bell System*, 104–106.

73 **the audiometer:** Fletcher, *Speech and Hearing*, 211–221.

73 **created the decibel:** Ibid., 68–69.

73 **20 to 20,000 Hz:** Ibid., 144.

73 **whisper to a yell:** Ibid., 69.

73 **Thirty-Fourth Street and Sixth Avenue:** Ibid., 106.

73 **1939 World's Fair:** Ibid., p. 97.

74 **Alfred I. duPont:** Joseph Frazier Wall, *Alfred I. DuPont: The Man and His Family* (New York: Oxford University Press, 1990), 109.

74 **According to Fletcher:** Harvey Fletcher 1884–1981, 175–176, and Bogert interview.

75 **"the first hearing aid":** Bogert interview.

75 **Fletcher made hearing aids for Thomas Edison:** Fletcher, *Harvey Fletcher 1884–1981*, 176–178.

76 **Békésy's traveling wave:** Author interview with Andrew Oxenham; Jürgen Tonndorf, "Georg von Békésy and his Work," *Hearing Research* 22 no.1–3 (1986): 3–10; Rossing et al., *Science of Sound*, 85-86; "Sound from Silence: The Development of Cochlear Implants," *Beyond Discovery*, National Academy of Sciences.

76 **After World War II:** Tonndorf, "Georg von Békésy": 4.

77 **"This space-time pattern":** Peter Dallos and Barbara Canlon, "Introduction to 'Good Vibrations': A Special Issue to Celebrate the 50th Anniversary of the Nobel Prize to Georg von Békésy," *Hearing Research* 293 no. 1–2 (2012): 1–2.

CHAPTER 7: WORD BY WORD

80 **Speech production:** Rossing et al., *Science of Sound*, 337–352; Author interview with David Poeppel.

86 **causes of hearing loss:** Waldman and Roush, *Your Child's Hearing Loss*, 29; Mark Almond and David J. Brown, "The Pathology and Etiology of Sensorineural Hearing Loss and Implications for Cochlear Implantation," in Niparko, ed., *Cochlear Implants: Principles & Practices*, 2nd ed., 43–81; Brad A. Stach and Virginia S. Ramachandran, "Hearing Disorders in Children," and Heidi L. Rehm and Rebecca Madore, "Genetics of Hearing Loss," in Jane R. Madell and Carol Flexer, eds., *Pediatric Audiology: Diagnosis, Technology, and Management* (New York: Thieme, 2008), 3–24; American Speech-Language-Hearing Association, at http://www.asha.org/PRPSpecificTopic.aspx?folderid=8589934680&section=Causes.

86 **CT scan:** *Medical News Today*, June 10, 2009, at http://www.medicalnewstoday.com/articles/153201.php.

87 **Mondini dysplasia or Mondini deformity:** Niparko, ed., *Cochlear Implants: Principles & Practices*, 49; National Institutes of Health Office of Rare Diseases Research, at http://rarediseases.info.nih.gov/gard/8215/mondini-dysplasia/resources/1.

87 **It was rare:** The definition of a rare disease is one that affects fewer than 200,000 people nationally: http://rarediseases.info.nih.gov/about-ordr/pages/31/frequently-asked-questions.

87 **EVA:** http://www.nidcd.nih.gov/health/hearing/pages/eva.aspx.

CHAPTER 8: THE HUB

This chapter is based on author interviews with Bill House, John House, David House, Michael Merzenich, Don Eddington, Marc Eisen, John Niparko, Simon Parisier, Mario Svirsky, Michael Dorman, and Paulette Fiedor.

90 **roundly criticized:** William F. House and J. Urban, "Long Term Results of Electrode Implantation and Electronic Stimulation of the Cochlea in Man," *The Annals of Otology, Rhinology, and Laryngology* 82 no. 4 (1973): 504–517.

90 **"Otology needs a new surgery":** William F. House, *The Struggles of a Medical Innovator: Cochlear Implants and Other Ear Surgeries*, (CreateSpace, 2011), 77–78; Richard T. Miyamoto, Marion Downs Lecture in Pediatric Audiology, AudiologyNOW, Dallas, April 2009.

90 **"If I tell you":** William F. House, *Cochlear Implants: My Perspective*, (Newport Beach, CA: AllHear, 1995), 5.

90–91 **Luigi Galvani** and **Alessandro Volta:** Saurabh B. Shah, Jeannie H. Chung, and Robert K. Jackler, "Lodestones, Quackery, and Science: Electrical Stimulation of the Ear Before Cochlear Implants," *American Journal of Otology* 18 no. 5 (1997): 665–670.

92 **"I received a shock in the head":** Ibid., 666; "Luigi Galvani," *Encyclopaedia Britannica*, at http://www.britannica.com/EBchecked/topic/224653/Luigi-Galvani;

"Conte Alessandro Volta," *Encyclopaedia Britannica*, at http://www.britan nica.com/EBchecked/topic/632433/Conte-Alessandro-Volta.

92 **André Djourno** and **Charles Eyriès:** Marc D. Eisen, "Djourno, Eyries, and the First Implanted Electrical Neural Stimulator to Restore Hearing," *Otology and Neurology* 24 no. 3 (2003): 500–506; Phillip R. Seitz, "French Origins of the Cochlear Implant," *Cochlear Implants International* 3 no. 2 (2002), 77–86; Stuart Blume, *The Artificial Ear: Cochlear Implants and the Culture of Deafness* (Piscataway, NJ: Rutgers University Press, 2010), 30–32; **expressed the desire**, 31.

94–95 **When acoustic energy is naturally translated . . . No one was sure:** "Sound from Silence"; Donald K. Eddington and Michael L. Pierschalla, "Cochlear Implants: Restoring Hearing to the Deaf," *On the Brain* (Harvard Mahoney Neuroscience Institute *Newsletter*) 3 no. 4 (1994); Donald K. Eddington, "Speech Recognition in Deaf Subjects with Multichannel Intracochlear Electrodes," *Annals of the New York Academy of Sciences* 405 no. 1 (1983): 241–258; F. Blair Simmons, "Cochlear Implants," *Archives of Otolaryngology* 89 no. 1 (1969): 61–69.

95 **"The more a researcher knew":** Blake S. Wilson and Michael F. Dorman, *Better Hearing with Cochlear Implants: Studies at the Research Triangle Institute* (San Diego, CA: Plural Publishing, 2012), 1.

95 **Bill House:** Author interview and House, *Struggles of a Medical Innovator* (**I could feel the joy**, 5; **from 40 percent to less than 1 percent**, 60; **Alan Shepard**, 49–50; **Jim Doyle**, 67–68).

98 **"Electronic Firm Restores Hearing with Transistorized System in Ear":** *Space Age News* 3 no. 21 (1961).

99 **Blair Simmons:** Blume, *Artificial Ear*, 32–38 (**"irresponsible claims" . . . "We were amazingly lucky,"** 33). W. E. Fee Jr. and Richard L. Goode, "F. Blair Simmons MD (1930–1998)," *Archives of Otolaryngology Head and Neck Surgery* 124 no. 8 (1998): 843–844; and W. E. Fee Jr. and Richard L. Goode, "Memorial Resolution: F. Blair Simmons (1930–1988)," online at http://histsoc.stanford.edu/pdf-mem/SimmonsFB.pdf.

99 **An eighteen-year-old cancer patient:** F. Blair Simmons et al., "Electrical Stimulation of Acoustical Nerve and Inferior Colliculus," *Archives of Otolaryngology Head and Neck Surgery* 79 (1964): 559–567.

99 **Anthony Vierra of San Jose:** F. Blair Simmons et al., "Auditory Nerve: Electrical Stimulation in Man," *Science* 148 no. 3666 (April 2, 1965): 104–106; F. Blair Simmons, "A History of Cochlear Implants in the United States: A Personal Perspective," in R. A. Schindler and M. M. Merzenich, eds., *Cochlear Implants*, (San Diego, CA: Raven, 1985), 1–7.

100 **TELL US WHEN:** Michael Dorman showed me a photograph of this sign.

101–102 **I am glad this meeting is a workshop . . . just _might_ be possible:** Simmons, "Cochlear Implants" (**"unabashed admiration . . . auto horns,"** [transcript] is in "Discussion" section).

102 **Charles "Chuck" Graser:** Mara Mills, "Do Signals Have Politics? Inscribing Abilities in Cochlear Implants," in Trevor Pinch and Karin Bijsterveld, eds., _The Oxford Handbook of Sound Studies_, (New York: Oxford University Press, 2011), 320–346 (**writing to House,** 329; **"You would probably describe,"** 330). _The Electronic Ear_, film by Karen House (sent to me by Paulette Fiedor).

105 **Parts of the movie:** House, _Struggles of a Medical Innovator_, 79. The National Geographic special aired in 1975. http://en.wikipedia.org/wiki/The_Incredible_Machine_(film).

106 **"Enthusiastic testimonials":** Pinch and Bijsterveld, eds., _Oxford Handbook of Sound Studies_, 331.

107 **Wright brothers' flight:** House, _Struggles of a Medical Innovator_, 81.

107 **Bilger report:** R. C. Bilger et al., "Evaluation of Subjects Presently Fitted with Implanted Auditory Prostheses," _Annals of Otology, Rhinology, and Laryngology_ 86 Supplement 38 (1977): 3–10; _Nashua Telegraph_, December 7, 1976, at http://news.google.com/newspapers?nid=2209&dat=19761207&id=UKYrAAAAIBAJ&sjid=3_wFAAAAIBAJ&pg=7066,1296511; House, _Cochlear Implants: My Perspective_, 8–9; Eddington, "Speech Recognition in Deaf Subjects."

CHAPTER 9: PRIDE

This chapter makes use of author interviews with Ted Supalla and Carol Padden.

109 **National Theatre of the Deaf:** Jack R. Gannon, _Deaf Heritage: A Narrative History of Deaf America_ (Silver Spring, MD: National Association of the Deaf, 1981). (**"He bowed slightly"**/**"As a result, sign language"**/**"They decided to make NTD,"** _Boston Herald_ and _The National Observer_, 346; _My Third Eye_ transcript, 354–355); Stephen C. Baldwin, _Pictures in the Air: The Story of the National Theatre of the Deaf_ (Washington, DC: Gallaudet University Press, 1993) (**a theatrical version of signed English**, 34); National Theatre of the Deaf website at http://www.ntd.org/about.php?id=history (**six people bought tickets**); Padden and Humphries, _Deaf in America_, chap. 5.

112 **William Stokoe:** Padden and Humphries, _Deaf in America_, chap. 5, particularly pp. 79–81; Sacks, _Seeing Voices_, 61–63.

113 **"He was the first linguist":** Gannon, _Deaf Heritage_, 364–367.

114 **"a study in the anatomy of prejudice":** Lane, _When the Mind Hears_, xiii, xv.

115 **"As I signed":** Lou Ann Walker, _A Loss for Words: The Story of Deafness in a Family_ (New York, Harper Perennial, 1987), 202.

115 **"became more self-conscious":** Padden and Humphries, *Inside Deaf Culture*, 130.

115 **"A large population":** *Deaf in America*, 9.

115–116 **"The traditional way ... to portray":** Padden and Humphries, *Deaf in America*, 1.

116 **not something to be "cured" or "fixed":** Dolnick, *Atlantic Monthly*.

116 **Deaf President Now:** Jack R. Gannon. *The Week the World Heard Gallaudet* (Washington, DC: Gallaudet University Press, 1989) (WE STILL HAVE A DREAM! p. 109 and frontispiece); Sacks, *Seeing Voices*, 99–130; Gallaudet University website, at http://www.gallaudet.edu/dpn_home.html; "Person of the Week," ABC News, March 11, 1988; "New President Protested at School for Deaf," *New York Times*, March 7, 1988; "Gallaudet University Installs Deaf President," *New York Times*, October 23, 1988.

CHAPTER 10: LANGUAGE IN THE BRAIN

This chapter makes use of author interviews with Simon Parisier, Michael Merzenich, David Poeppel, and Helen Neville.

123–126 **The differences between a brain ... where it is most desirable ... brain's basic division of labor:** *Changing Brains: Effects of Experience on Human Brain Development* (DVD), University of Oregon Brain Development Lab, 2009, at www.changingbrains.org; Lise Eliot, *What's Going on in There? How the Brain and Mind Develop in the First Five Years of Life* (New York: Bantam, 1999); Sharon Begley, "Your Child's Brain," *Newsweek*, February 18, 1996; Arthur S. Bard and Mitchell G. Bard, *The Complete Idiot's Guide to Understanding the Brain* (New York: Alpha Books, 2002).

124 **"merely an opening gambit":** Sharon Begley, *Train Your Mind, Change Your Brain: How a New Science Reveals Our Extraordinary Potential to Transform Ourselves* (New York: Ballantine Books, 2007), 10.

125 **"experience is like a sculptor":** *Changing Brains* DVD.

125 **National Institutes of Health and UCLA:** Nitin Gogtay et al., "Dynamic Mapping of Human Cortical Development during Childhood Through Early Adulthood," *PNAS* 101 no. 21 (2004): 8174–8179. See video at http://www.loni.ucla.edu/~thompson/DEVEL/dynamic.html.

126 **a second, smaller wave of neuron creation:** Jay N. Giedd et al., "Brain Development During Childhood and Adolescence: A Longitudinal MRI Study," *Nature Neuroscience* 2 no. 10 (1999): 861–863.

126 **possible to keep learning:** Begley, *Train Your Mind*, 28.

126 **The central auditory system:** Author interviews with Michael Merzenich and David Poeppel. Also, Bradford J. May and John K. Niparko,

"Auditory Physiology and Perception," in Niparko, ed., *Cochlear Implants: Principles & Practices,* 1–17; biographical monograph, Michael Merzenich, courtesy of Michael Merzenich.

129 **"If a system can be influenced":** Author interview with Helen Neville. See also "Nature and Nurture and the Developing Brain," Oregon Health and Science University Brain Awareness presentation by Neville, transcript courtesy of Neville Lab.

129 **A study published in 2009:** *USA Today,* December 7, 2008; See also http://www.pleasanton.k12.ca.us/avhsweb/emersond/appsych/ch3_brain/ses_brain.pdf; Rajeev D. S. Raizada and Mark M. Kishiyama, "Effects of Socioeconomic Status on Brain Development, and How Cognitive Neuroscience May Contribute to Levelling the Playing Field," *Frontiers in Human Neuroscience* 4 no. 3 (2010).

CHAPTER 11: WHAT IF THE BLIND COULD SEE?

This chapter is based on interviews with Helen Neville, Daphne Bavelier, Anu Sharma, Michael Dorman, and Michael Merzenich.

130–131 **Molyneux and Locke** and **Santiago Ramón y Cajal:** Begley, *Train Your Mind,* 30–37, 91–104 **("fixed, ended, immutable,"** p. 36).

131 **Hubel and Wiesel:** David H. Hubel and Torsten N. Wiesel, *Brain and Visual Perception: The Story of a 25-Year Collaboration* (New York: Oxford University Press, 2005) (**"That is something,"** p. 371); Robert H. Wurtz, "Recounting the Impact of Hubel and Wiesel," *Journal of Physiology* 587 no. 12 (2009): 2817–2823.

133 **Helen Neville:** Author Interview. Begley, *Train Your Mind,* 91–104 (**"which is where any well-behaved brain,"** p. 99); Daphne Bavelier and Helen J. Neville, "Cross-modal Plasticity: Where and How?" *Nature Reviews Neuroscience* 3 no. 6 (2002): 443–452; Helen Neville and Daphne Bavelier, "Human Brain Plasticity: Evidence from Sensory Deprivation and Altered Language Experience," *Plasticity in the Adult Brain: From Genes to Neurotherapy* 138 (2002): 177; Helen J. Neville and Daphne Bavelier, "Effects of Auditory and Visual Deprivation on Human Brain Development," *Clinical Neuroscience Research* 1 no. 4 (2001): 248–257.

137 **Anu Sharma:** Author interview; Anu Sharma, Michael F. Dorman, and Anthony J. Spahr, "A Sensitive Period for the Development of the Central Auditory System in Children with Cochlear Implants: Implications for Age of Implantation," *Ear and Hearing* 23 no. 6 (2002): 532–539; Phillip M. Gilley, Anu Sharma, and Michael F. Dorman, "Cortical Reorganization in Children with Cochlear Implants," *Brain Research* 1239 (2008): 56–65; Anu

Sharma, Amy A. Nash, and Michael Dorman, "Cortical Development, Plasticity and Re-organization in Children with Cochlear Implants," *Journal of Communication Disorders* 42 no. 4 (2009): 272–279.

CHAPTER 12: CRITICAL BANDWIDTHS

This chapter is based on interviews with Graeme Clark, Margaret Clark, Richard Dowell, Hugh McDermott, Peter Blamey, Rob Shepherd, Bob Cowan, Michael Merzenich, Don Eddington, and Michael Dorman.

140 **Graeme Clark:** Author interviews; Graeme Clark, *Sounds from Silence: Graeme Clark and the Bionic Ear Story*, (Crows Nest, NSW, Australia: Allen & Unwin, 2000) (**a colleague joked**, 211; **one particularly stubborn surgical question**, 93; **original diagram wider and higher**, 86; **Rod Saunders**, 117–121; quote from **George Watson**, 108; **F0F2**, 195–204); Blume, *The Artificial Ear*, 42–57.

144 **320,000 people use them:** Because some of the manufacturers of cochlear implants are private companies, it is difficult to get an accurate, current figure for the number of users worldwide. This is the most recent figure available. It was provided by the Lasker Foundation when they awarded the 2013 Lasker-DeBakey Clinical Medical Research Award to Graeme Clark, Ingeborg Hochmair, and Blake Wilson.

150 **UCSF:** Author interviews; Robin P. Michelson, "Cochlear Implants: Personal Perspectives," in Robert A. Schindler and Michael M. Merzenich, eds., *Cochlear Implants* (New York: Raven, 1985).

151 **"Our system extracted information":** Biographical monograph, Michael Merzenich, courtesy of Michael Merzenich; Blume, *The Artificial Ear*, 37, 46.

154 **Fletcher had introduced the concept of critical bandwidth:** Fletcher interview with Bruce Bogert at Bell Labs circa 1963; Millman, ed., *A History of Engineering and Science in the Bell System*, 99–103 (vocoder).

155 **University of Utah:** Author interviews; *On the Brain*, Fall 1994; Eddington, "Speech Recognition in Deaf Subjects."

157 **FDA approval:** Blume, *The Artificial Ear*, 46–57.

158 **The Utah device was sold:** Author interview with Don Eddington.

158 **"for the first time":** FDA deputy commissioner Mark Novitch; Douglas Martin, "Dr. William F. House, Inventor of Pioneering Ear-Implant Device, Dies at 89," *New York Times*, December 15, 2012.

CHAPTER 13: SURGERY

160 **I know generally how the surgery went:** Cochlear implant surgery is described in Debara L. Tucci and Thomas M. Pilkington, "Medical and Surgical Aspects of Cochlear Implantation," in Niparko, ed., *Cochlear Implants: Principles*

& *Practices*, 161–186; video demonstration at http://www.youtube.com/watch?v=WMe3yr2ZnUI; George Alexiades et al., "Cochlear Implants for Infants and Children," in Madell and Flexer, eds., *Pediatric Audiology*, 183–191.

CHAPTER 14: FLIPPING THE SWITCH

This chapter is based on personal recollection as well as a later interview with Lisa Goldin, and George Alexiades et al., "Cochlear Implants for Infants and Children," in Madell and Flexer, eds., *Pediatric Audiology*.

169 **Daniel Ling:** For Six-Sound Test, see Jane R. Madell, "Evaluation of Speech Perception in Infants and Children," in Madell and Flexer, eds., *Pediatric Audiology*, 91.

171 **"bimodal" hearing:** See R. H. Gifford et al., "Combined Electric and Contralateral Acoustic Hearing: Word and Sentence Recognition with Bimodal Hearing," *Journal of Speech, Language, and Hearing Research* 50 no. 4 (2007): 835; Ting Zhang, Michael F. Dorman, and Anthony J. Spahr, "Information from the Voice Fundamental Frequency (F0) Region Accounts for the Majority of the Benefit When Acoustic Stimulation Is Added to Electric Stimulation," *Ear and Hearing* 31 no. 1 (2010): 63–69; T. Y. C. Ching, E. Van Wanrooy, and H. Dillon, "Binaural-Bimodal Fitting or Bilateral Implantation for Managing Severe to Profound Deafness: A Review," *Trends in Amplification* 11 no. 3 (2007): 161–192.

CHAPTER 15: A PERFECT STORM

This chapter includes material from author interviews with Graeme Clark, Richard Dowell, Rob Shepherd, Peter Blamey, Hugh McDermott, Bill House, Paulette Fiedor, Simon Parisier, and Mark, Debi, and David Leekoff.

173 **Caitlin Parton:** *60 Minutes*, November 8, 1992.

173–177 **"People speak of the grief . . ." "There were no other families":** Melody James, speaking at the 102nd annual meeting of the Center for Hearing and Communication, online at http://www.chchearing.org/news-events/news-announcements/CHC-President-Letter%20.

173 **Fewer than three thousand people:** Wilson and Dorman, *Better Hearing with Cochlear Implants*, 2.

174 **"There is no moral justification":** *Medical World News*, June 11, 1984, 34.

175 **Tracy Husted:** See video at http://www.youtube.com/watch?v=dWJNX EeE0Vw; House, *Struggles of a Medical Innovator*, 88.

175 **Consensus was growing:** Wilson and Dorman, *Better Hearing with Cochlear Implants*, 2.

176 **Cochlear, the Australian company:** Clark, *Sounds from Silence*, chap. 11.

176 **Noel Cohen:** Ibid., 158, 184–185.

176 **the FDA had specified:** Ibid., p. 176.

176 **"They said this device":** Steve Parton on *60 Minutes*.

177–178 **a new round of protests . . . "the best time to be Deaf":** Andrew Solomon, "Defiantly Deaf," *New York Times*, August 28, 1994.

177 **"The deaf community has begun":** Dolnick, *Atlantic Monthly*, September 1993.

178 **Americans with Disabilities Act:** http://www.ada.gov/.

178–179 **"leveled the playing field" . . . "$2.5 billion per year":** Bonnie Poitras Tucker, "Deaf Culture, Cochlear Implants, and Elective Disability," *Hastings Center Report* 28 no. 4 (1998): 6–14.

179 **Cheryl Heppner told *The Atlantic Monthly*:** Dolnick, *Atlantic Monthly*, September 1993.

179 **1990 decision to approve cochlear implants:** Blume, *The Artificial Ear*, 55.

179 **just another in a long line of medical "fixes":** Author interviews with Ted Supalla, Carol Padden, Peter Hauser, Irene Leigh, and Matthew Bakke; Christiansen and Leigh, *Cochlear Implants in Children*, 256–161; Blume, *The Artificial Ear*, chap. 3; Padden and Humphries, *Inside Deaf Culture*, 166–168.

180 **"An implant is the ultimate invasion":** Matthew S. Moore and Linda Levitan, *For Hearing People Only*, 2nd ed., (Rochester, NY: Deaf Life, 1993), 191.

182 **Roslyn Rosen:** Quoted in *60 Minutes*.

182 **NAD had released an official statement:** National Association of the Deaf position papers on cochlear implants, 1991, 2000. See also Christiansen and Leigh, *Cochlear Implants in Children*, 257.

183 ***60 Minutes* also received angry letters:** *Atlantic Monthly*, 43.

183 **Confrontations between the Deaf community:** For Sourds en Colère see Blume, *The Artificial Ear*, 106; "Cochlear Implants Protest in France," DEAF-INFO, at http://www.zak.co.il/d/deaf-info/old/ci-france.

183 **increased risk of meningitis:** Raylene Paludneviciene and Raychelle L. Harris, "Impact of Cochlear Implants on the Deaf community," in Raylene Paludneviciene and Irene W. Leigh, eds., *Cochlear Implants: Evolving Perspectives* (Washington, DC, Gallaudet University Press, 2011), 6–7; Joseph Michael Valente, Benjamin Bahan, and H-Dirksen Bauman, "Sensory Politics and the Cochlear Implant Debates," in Paludneviciene and Leigh, eds., *Cochlear Implants: Evolving Perspectives*, chap. 12; Deaf Liberation Front press release, online at http://tech.dir.groups.yahoo.com/group/Bioethics/message/5594.

184 **In Melbourne, protesters:** Clark, *Sounds from Silence*, 184 (photo).

185 **"Most of us would *love*":** Tucker, "Deaf Culture, Cochlear Implants, and Elective Disability."

185 **the Smithsonian . . . was planning an exhibit:** http://www.handsandvoices .org/about/story.htm.

186 **Gallaudet University was in the news again:** Diana Jean Schemo, "Protests Continue at University for Deaf," *New York Times,* May 13, 2006; Jane K. Fernandes, "Many Ways of Being Deaf," *Washington Post,* October 14, 2006; Diana Jean Schemo, "Turmoil at College for Deaf Reflects Broader Debate, *New York Times,* October 21, 2006 (**"More parents are choosing"**); John B. Christiansen, *Reflections: My Life in the Deaf and Hearing Worlds* (Washington, DC: Gallaudet University Press, 2010), part 3.

188 **Middle States Commission on Higher Education:** "Statement of Accreditation Status: Gallaudet University," online at http://www.msche.org/documents/ SAS/237/Statement%20of%20Accreditation%20Status.htm.

188 **The numbers spoke for themselves:** Gallaudet University Annual Report, 2011.

CHAPTER 16: A CASCADE OF RESPONSES

This chapter is based on interviews with David Poeppel, Greg Hickok, Andrew Oxenham, Usha Goswami, and Jeff Walker and visits to the Poeppel Lab at New York University.

191 **"run, don't walk":** Poeppel speaking at the McGovern Institute, April 27, 2012, online at http://video.mit.edu/watch/2012-mcgovern-institute-symposium -david-poeppel-11234/.

191 **MEG:** See http://web.mit.edu/kitmitmeg/whatis.html.

194 **Anu Sharma:** See chapter 11.

196 **Eric Kandel:** *The Leonard Lopate Show,* WNYC, April 16, 2013.

197 **Camillo Golgi and Santiago Ramón y Cajal, who shared a Nobel Prize:** http://www.nobelprize.org/nobel_prizes/medicine/laureates/1906/.

198 **Each point where the response:** The responses to stimuli are known as event-related potentials (ERPs). For more on ERPs, see Alexandra P. Fonaryova Key, Guy O. Dove, and Mandy J. Maguire, "Linking Brainwaves to the Brain: An ERP Primer," *Developmental Neuropsychology* 27 no. 2 (2005): 183–215.

200 **bottom-up processing:** For more information, see Andreas K. Engel, Pascal Fries, and Wolf Singer, "Dynamic Predictions: Oscillations and Synchrony in Top-down Processing," *Nature Reviews Neuroscience* 2 no. 10 (2001): 704–716.

201 **Alvin Liberman . . . motor theory of speech perception:** Alvin M. Liberman and Ignatius G. Mattingly, "The Motor Theory of Speech Perception Revised," *Cognition* 21 no. 1 (1985): 1–36.

201 **representations and computations:** See Patricia Smith Churchland and Terrence J. Sejnowski, "Neural Representation and Neural Computation,"

Philosophical Perspectives 4 (1990): 343–382; David C. Knill and Alexandre Pouget, "The Bayesian Brain: The Role of Uncertainty in Neural Coding and Computation," *TRENDS in Neurosciences* 27 no. 12 (2004): 712–719; Edmund T. Rolls and Gustavo Deco, *Computational Neuroscience of Vision* (New York: Oxford University Press, 2002); Eric I. Knudsen, Sascha du Lac, and Steven D. Esterly, "Computational Maps in the Brain," *Annual Review of Neuroscience* 10 no. 1 (1987): 41–65.

202 **predictive coding:** For an example of how predictive coding works, see Gregory Hickok, "The Cortical Organization of Speech Processing: Feedback Control and Predictive Coding the Context of a Dual-Stream Model," *Journal of Communication Disorders* 45 no. 6 (2012): 393–402.

203 **Andrew Oxenham:** For examples of Andrew Oxenham's work on cochlear implants, see Michael K. Qin and Andrew J. Oxenham, "Effects of Simulated Cochlear-Implant Processing on Speech Reception in Fluctuating Maskers," *The Journal of the Acoustical Society of America* 114 no. 1(2003): 446–454; Andrew J. Oxenham, "Pitch Perception and Auditory Stream Segregation: Implications for Hearing Loss and Cochlear Implants," *Trends in Amplification* 12 no. 4 (2008): 316–331.

204 **sine waves:** Robert E. Remez et al., "Speech Perception Without Traditional Speech Cues," *Science* 212 no. 4497 (1981): 947–950; see also http://www.haskins.yale.edu/research/sws.html.

205 **Bob Shannon:** For Shannon's study, see Robert V. Shannon et al., "Speech Recognition with Primarily Temporal Cues," *Science* 270 no. 5234 (1995): 303–304.

206 **Ghitza and Greenberg:** Oded Ghitza and Steven Greenberg, "On the Possible Role of Brain Rhythms in Speech Perception: Intelligibility of Time-Compressed Speech with Periodic and Aperiodic Insertions of Silence," *Phonetica* 66 no. 1–2 (2009): 113–126.

CHAPTER 17: SUCCESS!

For personal chapters, I relied on my journals, recollections of events, and files of reports about Alex from audiologists, doctors, speech therapists, and teachers. I also interviewed some of the professionals who have worked with Alex over the years.

CHAPTER 18: THE SEARCH FOR EVIDENCE

This chapter uses material from interviews with Don Eddington, Michael Dorman, Elissa Newport, Mario Svirsky, Paulette Fiedor, the Leekoff family, Peter Hauser, Marc Marschark, David Pisoni, and Daphne Bavelier.

217 **In 1987 . . . The conclusion from that year:** Wilson and Dorman, *Better Hearing with Cochlear Implants*, 2.

217 **a far more effective speech processing program:** Author interviews with Michael Dorman and Don Eddington; Wilson and Dorman, *Better Hearing with Cochlear Implants,* 11–20.

219 **parents didn't have to be all that good:** Jenny L. Singleton and Elissa L. Newport, "When Learners Surpass Their Models: The Acquisition of American Sign Language from Inconsistent Input," *Cognitive Psychology* 49 no. 4 (2004): 370–407; Danielle S. Ross and Elissa L. Newport, "The Development of Language from Non-native Linguistic Input," *Proceedings of the 20th Annual Boston University Conference on Language Development* 2 (1996): 634–645.

221 **four portraits of Abraham Lincoln:** Mario A. Svirsky et al., "Current and Planned Cochlear Implant Research at New York University Laboratory for Translational Auditory Research," *Journal of the American Academy of Audiology* 23 no. 6 (2012): 422–437.

223 **enormous variability:** Marschark, *Raising and Educating a Deaf Child,* 49–56.

224 **"Kids with implants are doing better":** Marschark writing on website for *Raising and Educating a Deaf Child,* June 21, 2011, at http://www.rit.edu/ntid/educatingdeafchildren/?cat=4&paged=3.

224 **Mario Svirsky:** For examples of Svirsky's research on outcomes, see Mario A. Svirsky et al., "Language Development in Profoundly Deaf Children with Cochlear Implants," *Psychological Science* 11 no. 2 (2000): 153–158; Mario A. Svirsky et al., "Development of Language and Speech Perception in Congenitally, Profoundly Deaf Children as a Function of Age at Cochlear Implantation," *Audiology and Neurotology* 9 no. 4 (2004), 224–233.

   Niparko study: John K. Niparko et al., "Spoken Language Development in Children Following Cochlear Implantation," *JAMA: The Journal of the American Medical Association* 303 no. 15 (2010): 1498–1506.

   Geers study: Ann E. Geers, Johanna G. Nicholas, and Allison L. Sedey, "Language Skills of Children with Early Cochlear Implantation," *Ear and Hearing* 24 no. 1S (2003): 46S–58S; Emily A. Tobey et al., "Factors Associated with Development of Speech Production Skills in Children Implanted by Age Five," *Ear and Hearing* 24 no. 1S (2003): 36S–45S; Ann E. Geers, "Predictors of Reading Skill Development in Children with Early Cochlear Implantation," *Ear and Hearing* 24 no. 1S (2003): 59S–68S; Ann Geers et al., "Long-Term Outcomes of Cochlear Implantation in the Preschool Years: From Elementary Grades to High School," *International Journal of Audiology* 47 no. S2 (2008): S21–S30.

227 **"We felt retarded":** Jackie Roth quoted in Andrew Solomon, "Defiantly Deaf," *New York Times,* Aug. 28, 1994.

227 **the philosophy underlying deaf education had changed:** For the history of changes in deaf education, see Paludneviciene and Harris, "Impact of Cochlear Implants on the Deaf Community," and Harry G. Lang, "Perspectives on the

History of Deaf Education," in *The Oxford Handbook of Deaf Studies, Language, and Education*, vol. 1, 2nd ed. (New York: Oxford University Press, 2011); *Toward Equality: The Education of the Deaf*, 1988, at http://archive.gao.gov/t2pbat17/135760.pdf; in Marschark and Hauser, eds., *Deaf Cognition*, see Bavelier et al. on visual attention (chap. 9), Pisoni on cochlear implants (chap. 3), Marschark on language comprehension (chap.12), and Hauser on executive function (chap. 11).

228 **"total communication"** and **SimCom:** Marschark, *Raising and Educating a Deaf Child*, 66–67.

228 **"shouting":** Harlan Lane quoted in *Atlantic Monthly*, p. 50.

228 **Cued Speech:** Ibid., 88–89.

229 **bilingual-bicultural:** Ibid., 147–148.

229 **The question is no longer:** Marschark and Hauser, eds., *Deaf Cognition*, chap. 16.

229 **"First, there has never been any real evidence":** Marshark, *Raising and Educating a Deaf Child*, 4.

229 **"Effective parent-child communication":** Ibid., 5.

230 **"as a hearing person":** Ibid., 7.

232 **"We have to consider":** Marschark and Hauser, eds., *Deaf Cognition*, chap. 16.

235 **"Are there any deaf children":** Ibid.

CHAPTER 19: A PARTS LIST OF THE MIND

236 **Broca's area and Wernicke's area:** See Goldstein, *Sensation and Perception*, 8th ed., 323.

237 **Phineas Gage:** Steve Twomey, "Phineas Gage: Neuroscience's Most Famous Patient," *Smithsonian*, January 2010, online at http://www.smithsonianmag.com/history-archaeology/Phineas-Gage-Neurosciences-Most-Famous-Patient.html.

238 **a macaque monkey's visual system:** David C. Van Essen, Charles H. Anderson, and Daniel J. Felleman, "Information Processing in the Primate Visual System: An Integrated Systems Perspective." *Science* 255 no. 5043 (1992): 419–423.

238 **a map of the auditory system:** Jon H. Kaas, and Troy A. Hackett, "Subdivisions of Auditory Cortex and Processing Streams in Primates." Proceedings of the National Academy of Sciences 97 no. 22 (2000): 11793–11799.

239 **"I say to you 'cat'":** From David Poeppel presentation, 2013 AAAS Annual Meeting, Boston.

239 **Hickok and Poeppel's model:** Gregory Hickok and David Poeppel, "The Cortical Organization of Speech Processing," *Nature Reviews Neuroscience* 8 no. 5 (2007): 393–402; Gregory Hickok and David Poeppel, "Dorsal and Ventral Streams: A Framework for Understanding Aspects of the Functional Anatomy of Language," *Cognition* 92 no. 1 (2004): 67–99; Gregory Hickok and David

Poeppel, "Towards a Functional Neuroanatomy of Speech Perception," *Trends in Cognitive Sciences* 4 no. 4 (2000): 131–138.

241 **I even found it in a new textbook:** Goldstein, *Sensation and Perception*, 323.

242 **Each of those linguistic tasks:** Dorit Ben Shalom and David Poeppel, "Functional Anatomic Models of Language: Assembling the Pieces," *Neuroscientist* 14 no. 1 (2008): 119–127; Eric Pakulak and Helen Neville, "Biological Bases of Language Development," *Encyclopedia on Early Childhood Development*, Centre of Excellence for Early Childhood Development, published online April 28, 2010, at http://www.child-encyclopedia.com/pages/pdf/pakulak-nevilleangxp.pdf; Helen J. Neville and Daphne Bavelier, "Neural Organization and Plasticity of Language," *Current Opinion in Neurobiology* 8 no. 2 (1998): 254–258. For more on development of linguistic tasks and reading, see Maryanne Wolf, *Proust and the Squid: The Story and Science of the Reading Brain* (New York: HarperCollins, 2007), 113.

243 **Janet Werker:** For Werker study, see Judit Gervain and Janet F. Werker, "Prosody Cues Word Order in 7-Month-Old Bilingual Infants," *Nature Communications* 4 no. 1490 (2013).

245 **"word onset effect":** Helen Neville, "Nature and Nurture and the Developing Brain," talk for OHSU Brain Awareness.

245 **Usha Goswami:** For examples of Usha Goswami's work, see Usha Goswami et al., "Amplitude Envelope Onsets and Developmental Dyslexia: A New Hypothesis," *PNAS* 99 no. 16 (2002): 10911–10916; Jennifer M. Thomson and Usha Goswami, "Rhythmic Processing in Children with Developmental Dyslexia: Auditory and Motor Rhythms Link to Reading and Spelling," *Journal of Physiology–Paris* 102 no. 1 (2008): 120–129.

CHAPTER 20: A ROAD MAP OF PLASTICITY

This chapter is based on interviews with Helen Neville, Eric Pakulak, and Michael Merzenich.

247 **Head Start:** Neville et al., "Family-Based Training Program Improves Brain Function, Cognition, and Behavior in Lower Socioeconomic Status Preschoolers," www.pnas.org/cgi/doi/10.1073/pnas.1304437110.

247 **Mike Merzenich:** For Merzenich's neuroplasticity studies, see Begley, *Train Your Mind*, 37–45.

249 **"neuroeducation":** For more on neuroeducation, see Gary Stix, "How to Build a Better Learner," *Scientific American*, August 2011; Sheida Rabipour and Amir Raz, "Training the Brain: Fact and Fad in Cognitive and Behavioral Remediation," *Brain and Cognition* 79 (2012): 159–179; Usha Goswami, "Neuroscience and Education: From Research to Practice?" *Nature Reviews Neuroscience*, AOP,

published online April 12, 2006, at http://www.uni.edu/gabriele/page4/files/goswami002820060029-neuroscience-and-education.pdf.

250 **"profiles in plasticity":** Neville, "Nature and Nurture and the Developing Brain" (**"road map of plasticity"**); Courtney Stevens and Helen Neville, "Profiles of Development and Plasticity in Human Neurocognition," in Michael Gazzaniga, ed., *The Cognitive Neurosciences*, 4th ed., (Cambridge, MA: MIT Press, 2009), 165–181; Helen J. Neville and Daphne Bavelier, "Neural Organization and Plasticity of Language," *Current Opinion in Neurobiology* 8 no. 2 (1998): 254–258.

252 **the better babies are at responding:** Patricia K. Kuhl et al., "Phonetic Learning as a Pathway to Language: New Data and Native Language Magnet Theory Expanded (NLM-e)," *Philosophical Transactions of the Royal Society of London—Series B: Biological Sciences* 363 no. 1493 (2008): 979–1000; Discussed in Eric Pakulak and Helen Neville, "Biological Bases of Language Development," *Encyclopedia on Early Childhood Development*, 2010. http://www.child-encyclopedia.com/pages/PDF/Pakulak-NevilleANGxp.pdf.

252 **in thirteen-month-olds:** Debra L. Mills, Sharon Coffey-Corina, and Helen Neville, "Language Comprehension and Cerebral Specialization from 13 to 20 Months," *Developmental Neuropsychology* 13 no. 3 (1997): 397–445.

252 **By twenty months:** Debra L. Mills, Sharon Coffey-Corina, and Helen Neville, "Language Comprehension and Cerebral Specialization in 20-Month-Old Infants," *Journal of Cognitive Neuroscience* 5 no. 3 (1993): 317–334.

252 **English-Korean speakers:** Elissa L. Newport, "Maturational Constraints on Language Learning," *Cognitive Science* 14 no. 1 (1990): 11–28; Jacqueline S. Johnson and Elissa L. Newport, "Critical Period Effects in Second Language Learning: The Influence of Maturational State on the Acquisition of English as a Second Language," *Cognitive Psychology* 21 (1989): 60–99.

253 **Pakulak also studied Germans:** Eric Pakulak and Helen J. Neville, "Maturational Constraints on the Recruitment of Early Processes for Syntactic Processing," *Journal of Cognitive Neuroscience* 23 no. 10 (2011): 2752–2765.

254 **"level of proficiency":** Eric Pakulak and Helen J. Neville, "Proficiency Differences in Syntactic Processing of Monolingual Native Speakers Indexed by Event-related Potentials," *Journal of Cognitive Neuroscience* 22 no. 12 (2010): 2728–2744.

254 **depend in part on selective attention:** Courtney Stevens and Helen Neville, "Different Profiles of Neuroplasticity in Human Neurocognition," in S. Lipina and M. Sigman, eds., *Cognitive Neuroscience and Education*, in press. http://bdl.uoregon.edu/Publications/Lipina%20chapter%20in%20press.pdf.

254 **50 to 100 percent stronger:** Ibid.

257 **Using EEG, the children all underwent the same study:** Courtney Stevens, Brittni Lauinger, and Helen Neville, "Differences in the Neural Mechanisms

of Selective Attention in Children from Different Socioeconomic Backgrounds: An Event-Related Brain Potential Study," *Developmental Science* 12 no. 4 (2009): 634–646; Lisa D. Sanders et al., "Selective Auditory Attention in 3- to 5-Year-Old Children: An Event-Related Potential Study," *Neuropsychologia* 44 no. 11 (2006): 2126–2138. See also Head Start study, above.

CHAPTER 21: "I CAN'T TALK!"

For personal chapters, I relied on my journals, recollections of events, and files of reports about Alex from audiologists, doctors, speech therapists, and teachers. I also interviewed some of the professionals who have worked with Alex over the years.

CHAPTER 22: THE READING BRAIN

This chapter is based on interviews with Ken Pugh, Steve Frost, Usha Goswami, David Poeppel, Marc Marschark, and Beth Benedict.

267 **"an optional accessory":** Steven Pinker, quoted in Wolf, *Proust and the Squid*, 19.

267–269 **a church of phonological awareness . . . At Haskins, the Libermans:** Donald Shankweiler, "Words to Meanings," *Scientific Studies of Reading* 3 no. 2 (1999): 13–127 (**Reversals of letters and words**); Bennett A. Shaywitz et al., "The Functional Organization of Brain for Reading and Reading Disability (Dyslexia)," *Neuroscientist* 2 no. 4 (1996): 245–255.

268 **"Made any discoveries today?":** Haskins Laboratories, *The Science of the Spoken and Written Word,* 2005. http://www.haskins.yale.edu/CaseState ment/Haskinscase.pdf.

268 **"the entire speech stream":** Wolf, *Proust and the Squid*, 66–68.

269 **preschool oral language skills are predictive:** Bruce Pennington, University of Denver, "Definitions and Comorbidities of Developmental Disorders," talk at 2011 Aspen Brain Forum.

269 **how the brain reacts to speech:** Bruce D. McCandliss, Vanderbilt University, "How Instructors Direct a Learner's Attention Impacts Neural Changes During Reading Acquisition," talk at 2011 Aspen Brain Forum.

269 **location of a kindergartner's brain response:** U. Maurer et al., "Coarse Neural Tuning for Print Peaks when Children Learn to Read," *Neuroimage* 33 no. 2 (2006): 749–758; Sanne van der Mark et al., "The Left Occipitotemporal System in Reading: Disruption of Focal fMRI Connectivity to Left Inferior Frontal and Inferior Parietal Language Areas in Children with Dyslexia," *Neuroimage* 54 no. 3 (2011): 2426–2436.

269 **National Reading Panel:** National Institute of Child Health and Human Development, *Teaching Children to Read: An Evidence-Based Assessment of the Scientific Research Literature on Reading and its Implications for Reading Instruction* (US. Government Printing Office, 2000).

270 **state of California:** Dehaene, *Reading in the Brain*, 200.

271 *Charlotte's Web:* Wolf, *Proust and the Squid*, 131.

271 **informed by neuroscience:** For neuroscience of reading, see Bradley L. Schlaggar and Bruce D. McCandliss, "Development of Neural Systems for Reading," *Annual Review of Neuroscience* 30 (2007): 457–503; Stanislas Dehaene, *Reading in the Brain: The New Science of How We Read* (New York: Viking, 2009), chap. 5; and Wolf, *Proust and the Squid*, Part 2.

272 **at the University of Oregon:** Yoshiko Yamada et al., "Emergence of the Neural Network for Reading in Five-Year-Old Beginning Readers of Different Levels of Pre-Literacy Abilities: An fMRI Study," *Neuroimage* 57 no. 3 (2011): 704–713.

273 **brain's ability to connect:** Ibid., 94.

274 **visual word form area:** Bruce D. McCandliss, Laurent Cohen, and Stanislas Dehaene, "The Visual Word Form Area: Expertise for Reading in the Fusiform Gyrus," *Trends in Cognitive Sciences* 7 no. 7 (2003): 293–299.

274 **The eyes impose constraints:** Dehane, *Reading in the Brain*, 16–17.

274 **Italian** and **Mandarin:** Ibid., 31–37.

274 **Portuguese villages:** Paulo Ventura et al., "The Locus of the Orthographic Consistency Effect in Auditory Word Recognition," *Language and Cognitive Processes* 19 no. 1 (2004): 57–95.

275 **nonsense words:** Interview with Ken Pugh. Mentioned in Paula Tallal's interview on Phonological Processing, with David Boulton, Children of the Code. http://www.childrenofthecode.org/interviews/tallal.htm.

275 **A second study:** Mark S. Seidenberg and Michael K. Tanenhaus, "Orthographic Effects on Rhyme Monitoring," *Journal of Experimental Psychology: Human Learning and Memory* 5 no. 6 (1979): 546–554.

275 **Dennis and Victoria Molfese:** For examples of work by the Molfeses, see Dennis L. Molfese, "Predicting Dyslexia at 8 Years of Age Using Neonatal Brain Responses," *Brain and Language* 72 no. 3 (2000): 238–245; Kimberly Andrews Espy et al., "Development of Auditory Event-Related Potentials in Young Children and Relations to Word-Level Reading Abilities at Age 8 Years," *Annals of Dyslexia* 54 no.1 (2004): 9–38.

276 **"about ninety-nine percent"** correct: Kirsten Weir, "Catching Reading Problems Early," *Monitor on Psychology* 42 no. 4 (2011): 46.

276 **Goswami's theory:** Usha Goswami et al., "Amplitude Envelope Onsets and Developmental Dyslexia: A New Hypothesis," *PNAS* 99 no. 16 (2002): 10911–10916.

277 **Anne-Lise Giraud:** Katia Lehongre et al., "Altered Low-Gamma Sampling in Auditory Cortex Accounts for the Three Main Facets of Dyslexia," *Neuron* 72 no. 6 (2011): 1080–1090.

277 **poetry:** Ibid., 97.

277 **Lynette Bradley and Peter Bryant:** Ibid., 100.

277 **"As they listen":** Ibid., 223.

278 **introduction of early hearing screenings:** Karl R. White, "Newborn Hearing Screening," in Madell and Flexer, eds., *Pediatric Audiology*, 31–42.

279 **Ken Pugh:** For examples of Ken Pugh's work and more on the neuroscience of reading, see Ken Pugh et al., "Cerebral Organization of Component Processes in Reading," Brain 119 (1996): 1221–1238; Ken Pugh et al., "Predicting Reading Performance from Neuroimaging Profiles: The Cerebral Basis of Phonological Effects in Printed Word Identification," *Journal of Experimental Psychology: Human Perception and Performance* 23 no. 2 (1997): 299–318; Rebecca Sandak et al., "The Neurobiological Basis of Skilled and Impaired Reading: Recent Findings and New Directions," *Scientific Studies of Reading* 8 no. 3 (2004): 273–292.

279 **Research with students at Gallaudet:** Vicki L. Hanson and Carol A. Fowler, "Phonological Coding in Word Reading: Evidence from Hearing and Deaf Readers," *Memory and Cognition* 15 no. 3 (1987): 199–207.

280 **emphasis on phonology is misplaced:** Rachel I. Mayberry, Alex A. del Giudice, and Amy M. Lieberman, "Reading Achievement in Relation to Phonological Coding and Awareness in Deaf Readers: A Meta-Analysis," *Journal of Deaf Studies and Deaf Education* 16 no. 2 (2011): 164–188; Rachel I. Mayberry, "When Timing Is Everything: Age of First-Language Acquisition Effects on Second-Language Learning," *Applied Psycholinguistics* 28 no. 3 (2007): 537–549.

282 **What limited data there is:** See Outcomes studies in chap. 18. Also see Caitlin M. Dillon, Kenneth de Jong, and David B. Pisoni, "Phonological Awareness, Reading Skills, and Vocabulary Knowledge in Children Who Use Cochlear Implants," *Journal of Deaf Studies and Deaf Education* 17 no. 2 (2012): 205–226; Marschark, *Raising and Educating a Deaf Child*, chap. 7.

CHAPTER 23: DEAF LIKE ME

283 **"When my work faltered":** Josh Swiller, "I Think I Hear You," *Washingtonian*, published online September 13 2010, at http://www.washingtonian.com/articles/people/i-think-i-hear-you/.

287 **"This is often the way":** For a discussion of Ursula Bellugi's work, see Sacks, *Seeing Voices*, 63–82.

287 **"essentially nothing was known. . . . They invited deaf people":** "Remembering Dr. Edward S. Klima," *Inside Salk* 3 (2009). http://www.salk.edu/inside salk/print.php?id=91.

289 **neurobiological foundations of sign language:** Sacks, *Seeing Voices*, 74–75; Daphne Bavelier, David P. Corina, and Helen J. Neville, "Brain and Language: A Perspective from Sign Language," *Neuron* 21 no.2 (1998): 275–278; Mairéad MacSweeney et al., "The Signing Brain: The Neurobiology of Sign Language," *Trends in Cognitive Sciences* 12 no. 11 (2008): 432–440; Aaron J. Newman et al., "A Critical Period for Right Hemisphere Recruitment in American Sign Language Processing," *Nature Neuroscience* 5 no. 1 (2002): 76–80; Helen J. Neville et al., "Neural Systems Mediating American Sign Language: Effects of Sensory Experience and Age of Acquisition," *Brain and Language* 57 no. 3 (1997): 285–308; Ruth Campbell, Mairéad MacSweeney, and Dafydd Waters, "Sign Language and the Brain: A Review," *Journal of Deaf Studies and Deaf Education* 13 no. 1 (2008): 3–20.

289 **"ASL is the only thing":** Barbara Kannapell, quoted in Sacks, *Seeing Voices*, 119.

290 **Reliable statistics:** "American Sign Language," *Ethnologue* 17th Edition, at https://www.ethnologue.com/language/ase/; "Languages Spoken at Home by Language: 2009," 2012 Statistical Abstract, United States Census Bureau, at http://www.census.gov/compendia/statab/2012/tables/12s0053.pdf; Ross E. Mitchell et al., "How Many People Use ASL in the United States? Why Estimates Need Updating," *Sign Language Studies* 6 no. 3 (2006): 306–335; Padden and Humphries, *Inside Deaf Culture*, 9.

291 **"The inherent capability of children to acquire ASL":** National Association of the Deaf on American Sign Language. http://nad.org/issues/american-sign-language.

291 **"What bilingualism does":** Ellen Bialystok speaking on *The Brian Lehrer Show*, WNYC, April 10, 2013.
  Research on bilingualism: Ellen Bialystok, Fergus I. M. Craik, and Gigi Luk, "Bilingualism: Consequences for Mind and Brain," *Trends in Cognitive Sciences* 16 no. 4 (2012): 240–250 (**"one of the chief factors"**); Yudhijit Bhattacharjee, "Why Bilinguals are Smarter," *New York Times*, March 17, 2012; Perri Klass, "Hearing Bilingual: How Babies Sort Out Language," *New York Times*, Oct. 10, 2011.

293 **"The fact that you're constantly manipulating":** Bialystok, *The Brian Lehrer Show*, WNYC, April 10, 2013.

CHAPTER 24: THE COCKTAIL PARTY PROBLEM

This chapter is based on author interviews with Michael Dorman, Sarah Cook, John Ayers, Don Eddington, Hugh McDermott, and Karyn Galvin.

296 **Lawrence Revit:** Described to me by Michael Dorman. For more information, see R-SPACE sound system, at http://www.revitronix.com/.

298 **Harvey Fletcher of Bell Labs described:** Fletcher and Arnold, *Speech and Hearing*, 99.

299 **understanding the cocktail party problem:** For a review of research on the cocktail party problem, see Christophe Micheyl and Andrew Oxenham, "Pitch, Harmonicity, and Concurrent Sound Segregation: Psychoacoustical and Neurophysiological Findings," *Hearing Research* 266 no. 1 (2010): 36–51.

300 **spatial localization:** Rossing et al., *Science of Sound*, 89–90; Goldstein, *Sensation and Perception*, 292–299.

301 **"temporal fine structure":** See Brian C. J. Moore, "The Role of Temporal Fine Structure Processing in Pitch Perception, Masking, and Speech Perception for Normal-Hearing and Hearing-Impaired People," *Journal of the Association for Research in Otolaryngology* 9 no. 4 (2008): 399–406.

302 **René Gifford:** "High Fidelity: Cochlear Implant Users Report Dramatically Better Hearing with New Vanderbilt Process," Vanderbilt University Medical Center *Reporter*, March 5, 2013. http://news.vanderbilt.edu/2013/03/high-fidelity/.

302 **Michael Dorman of Arizona State University:** For examples of Dorman's work, see M. F. Dorman et al., "Current Research with Cochlear Implants at Arizona State University," *Journal of the American Academy of Audiology* 23 no. 6 (2012): 385–395; Michael F. Dorman et al., "Speech Perception and Sound Localization by Adults with Bilateral Cochlear Implants," *Seminars in Hearing* 32 no. 1 (2011): 73–89.

303 **"bimodals":** For bimodal bilingualism, see Karen Emmorey et al., "The Source of Enhanced Cognitive Control in Bilinguals: Evidence from Bimodal Bilinguals," *Psychological Science* 19 no. 12 (2008): 1201–1206.

CHAPTER 25: BEETHOVEN'S NIGHTMARE

This chapter includes material from interviews with Charles Limb, Peter Blamey, Hugh McDermott, and Hamish Innes-Brown.

308 **marbles rolling around in a dryer:** Beverly Biderman, *Wired for Sound: A Journey into Hearing* (Toronto: Trifolium Books, 1998), 12.

309 **"Happy Birthday to You":** Patrick J. Donnelly and Charles J. Limb, "Music Perception in Cochlear Implant Users," in Niparko, ed., *Cochlear Implants: Principles & Practices*, 223–228.

310 **commissioned composers to create music:** The music commissioned for cochlear implant users was performed at an event called Interior Design: Music for the Bionic Ear, The Bionic Ear Institute, 2011. Hamish Innes-Brown described Natasha Anderson's music to me, and I read her description of it in the concert brochure.

311 *Through Deaf Eyes:* PBS, 2007.

311 **Dame Evelyn Glennie:** "Evelyn Glennie: How to Truly Listen," TED Talk, February 2003. http://www.ted.com/talks/evelyn_glennie_shows_how_to_listen.html.

312 **the first music that sounded good:** Biderman, *Wired for Sound*, 23–24; Donnelly and Limb, "Music Perception in Cochlear Implant Users."

312 **timbre:** For timbre perception in CI users, see Donnelly and Limb, "Music Perception," 226; Kate Gfeller et al., "Timbral Recognition and Appraisal by Adult Cochlear Implant Users and Normal-Hearing Adults," *Journal of the American Academy of Audiology* 9 no.1 (1998): 1–19.

CHAPTER 26: WALK BESIDE ME

319 **In 2000, the National Association of the Deaf revised its position:** See www.nad.org.

319 **half of cochlear implant recipients are children:** Julia Sarant, "Cochlear Implants in Children: A Review," in Sadaf Naz, ed., *Hearing Loss*, 2012. Available from http://www.intechopen.com/books/hearing-loss/cochlear-implants-in -children-a-review.

319 **In 2009, it was estimated:** Shari Roan, "Cochlear Implants Open Deaf Kids' Ears to the World," *Los Angeles Times*, August 3, 2009; John B. Christiansen and Irene W. Leigh, "Cochlear Implants and Deaf Community Perceptions," in Paludneviciene and Leigh, eds., *Cochlear Implants: Evolving Perspectives*, 52.

319 **Beginnings:** "10-Year Change in Communication Choice, 5/28/13," personal communication from Joni Alberg, executive director of Beginnings. The Beginnings DVD is available at https://www.ncbegin.org/#.

320 **an "uncomfortable" place:** Josh Swiller on his year at Gallaudet, in "I Think I Hear You."

326 **Gallaudet does look different:** For current statistics about Gallaudet University, see *Annual Report of Achievements, Fiscal Year 2012*, Gallaudet University. http://www.gallaudet.edu/Documents/Academic/FY%202012%20Annual% 20Report%20FINAL.pdf.

328 **"gradually giving way to a more nuanced view":** Paludneviciene and Leigh, *Cochlear Implants: Evolving Perspectives*, vii. See also Irene W. Leigh, *A Lens on Deaf Identities* (New York: Oxford University Press, 2009).

CHAPTER 27: FROG IN HONG KONG

For personal chapters, I relied on my journals, recollections of events, and files of reports about Alex from audiologists, doctors, speech therapists, and teachers. I also interviewed some of the professionals who have worked with Alex over the years.

warm, nurturing women who worked at our beloved Berkeley Carroll Child Care Center. Thank you to Barry Price and everyone at Premier Pediatrics, to Bernard and Linda, our first therapists, to Jay Dolitsky, George Alexiades, Jessica Lisogorsky, Tracey Vytlacil, Sabrina Vitulano, Katelyn Stoehr, Jillian Levine, Yvette, Melissa, and everyone at New York Eye and Ear Infirmary's Children's Hearing Institute, but especially Jessica O'Gara and Lisa Goldin, for their sustained care and attention, and Simon Parisier, for his wisdom and care not just for us but also for generations of families. Thank you Karen London for calling when I most needed to hear from you. Thank you to Teresa Boemio and everyone at Clarke New York, especially Melissa Arnott, Shari Rose, Janet Ellwood, Shauna Rogers, Alison Gregge, Anne Carney, and Lori Chalom. Thank you to Laura Reisler and everyone at Park Slope Communication Center. Thank you to David Spritzler and Karen Louick—we are so lucky to have you both working with Alex.

At Berkeley Carroll, there are so many people who make the school the stimulating and nurturing place that it is, and I am grateful to you all. Thank you particularly to Alex's teachers (through fourth grade): Norah Breen, Victoria Nachimson, Sherri Paller, Sarah Layden, Christine Eddis, Kate Keim, Rachel Cohen, Andrew Ahmadi, Mark Durbak, and all the specialists, especially Don Militello. And thank you to the administrators who always thought first about what was right for Alex: Ben Chant, Ellen Arana, Pam Cunningham, David Egolf, and Bob Vitalo. In Hong Kong, thank you to Riz Farooqi, Jill Kaufman, Gretchen Loughran, Bonnie Lim, Toni Bain, and Zoe Deane (boosh!).

Once I began to contemplate actually writing this book, one of the first people I talked to was John Niparko, whose graciousness and enthusiasm got me off to a good start—that it was also a slow one was entirely my fault. Not everyone I subsequently interviewed shows up in the manuscript, but each helped me frame the issues and understand things just a little better. Thank you to Nancy Mellon and the excellent people at the River School, David Ryugo, Charles Limb, Jane Madell,

Scott Stauffer, Elissa Newport, Ted Supalla, Daphne Bavelier, Peter Hauser, Carol Padden, Karen Emmorey, David Pisoni, and Marc Marschark. At AG Bell, thanks to Don Goldberg, Alexander Graham, Kathleen Treni, Susan Boswell, Garrett Yates, and Catherine McNally. Thanks to Mark Leekoff, and Debi and David Leekoff for sharing their story. I was fortunate and grateful to spend a day with Bill House, who passed away in December 2012. I thank him posthumously for his time and his trust. Thanks also to David House, John House, and Paulette Fiedor. Thanks to Mike Merzenich, whose interest in my project was so encouraging, and to Don Eddington and Paula Tallal. Thank you to Michael Dorman, Sarah Cook, John Ayers, and everyone in the Dorman lab. In Melbourne, heartfelt thanks to Graeme and Margaret Clark for making me feel so welcome, and to Rob Shepherd, Peter Blamey, Hugh McDermott, Julia Sarant, Karyn Galvin, Richard Dowell, Shani Dettman, Hamish Innes-Brown, and Nick Sinclair for giving me their time.

Thank you to Anu Sharma, Athena Vouloumanos, and everyone in their labs, to Janet Werker, Gina Lebedeva, and Erich Jarvis. Thank you to Helen Neville and everyone in the Neville lab, especially Eric Pakulak, who put up with me graciously for days and then withstood repeated follow-up questions. Thank you to Mario Svirsky and everyone in his lab at Bellevue, and to Greg Hickok, Andrew Oxenham, and Usha Goswami. At Haskins, thank you to Ken Pugh, Steve Frost, Doug Whalen, and Diane Lillo-Martin. Thank you to Jean Mary Zarate, Luc Arnal, Jeff Walker, and everyone in the Poeppel lab. A very special thank-you to David Poeppel, whose intelligence, sense of humor, and patience made the book far stronger than it would otherwise have been.

At Gallaudet, a heartfelt thank-you to Steve Weiner for welcoming me, to Irene Leigh, Angela McCaskill, Sam Swiller, Josh Swiller, Matthew Bakke, and Beth Benedict for being willing to talk, and to Catherine Murphy and Kaitlin Luna for their help. And a special thank-you to Janis Cole for extending a hand, so to speak. Thank you, too, to the ASL interpreters who worked with me at Gallaudet, Rochester, and Georgetown.

Quite a few of the people I interviewed graciously reviewed parts or all of the manuscript to confirm my facts and interpretations. Any remaining errors are mine.

Though it was challenging to research this book and master its various subjects, it was even harder to decide how best to frame the stories it tells and weave them together. I am grateful to Russ Galen for getting me started and to my friend Barbara Grossman, who gave me a fervent push just when I thought I might give up. Thanks to Jennifer Rudolph Walsh for connecting me with Dorian Karchmar, my extraordinary agent, who got it from the start and pulled the very best thinking and work out of me. Thank you, Dorian, for everything.

I am grateful to my editor, Stephen Morrow, who loved this book and cared about Alex from day one, whose enthusiasm was sustaining, and who was always willing to talk things through. Thanks also to Stephanie Hitchcock, Brian Tart (baseball coach extraordinaire), Liza Cassity, and everyone at Dutton, who believed so fervently in the book's potential. Thank you to Layla Lang for the illustrations.

No working mother goes it alone. Thanks to Marilyn Ling and family, Jacqueline Schuka, Anneli Lukk, Gillian O'Reilly, Katherina Zubryzcki, Ramona Freismuth, Yvonne Hoppe, Sean Donovan, Abby Browde, Henry Schwab, Pia Senoron, and Juliet Curammeng for taking such good care of the boys over the years. A heartfelt thank-you to Jessica Barthel for coming into our lives not once but twice, for loving Alex so very much, and for making it possible for me to have the space and the time to write this book.

Thank you to Moira Bailey for listening, for reading, for checking in, to Jenny McMahon for being such a faithful friend and sharing the agony and the ecstasy, to Amy and Tom Jakobson (and Linus and especially Nina!), and Elizabeth Schwarz (and the Firestone boys), who have all been there for me and Alex specifically and for the Jusths in general from the start, and to Stephanie Holmes, Allan Denis, Audrey Denis, and Christian Denis for making us feel like family. Thanks to Judy Warner and Max Berley for putting me up repeatedly in Washington. Thanks to Stella and Tony Biniaris for practically adopting Alex and to all of my Hong Kong friends for the welcome, the wine, and the walks. Back in Brooklyn, thanks to all the members of the Upswing Club, with whom I feel such a deep connection. You kept me going at critical junctures. Now I might actually have time for dinner.

Thanks to all the Denworths and Jusths for everything, and especially my mother, Joanne Denworth, for her constant support.

Jake, Matthew, and Alex: I am grateful for your love, your intelligence, and your laughter. You should know that I love being your home base. Jake and Matty, thank you for looking after Alex and teaching him the ways of the Justh brothers. Alex, thank you for displaying such courage, for inspiring me, and for letting me tell this story.

And finally, to my husband, Mark Justh, my partner in all things, thank you for your sense of humor, your passion, your generosity, and your belief in me and this book. You inspire us all to think big, and you've made life an adventure I feel lucky to have shared. Onward to the next challenge and the next twenty-five years!

INDEX